IMPURITY SPECTRA
OF SOLIDS

Elementary Theory of Vibrational Structure

IMPURITY SPECTRA
OF SOLIDS

Elementary Theory of Vibrational Structure

KARL K. REBANE

Institute of Physics and Astronomy
Academy of Sciences of the Estonian SSR
Tartu, USSR

Translated from Russian by
JOHN S. SHIER
NASA Electronics Research Center
Cambridge, Massachusetts

⨎ PLENUM PRESS • NEW YORK—LONDON • 1970

Karl Karlovich Rebane was born in 1926 in Pärnu in the Estonian SSR. He was graduated from Leningrad State University in 1952, and attained the degree of Doctor of Physicomathematical Science in 1964. His scientific work has been largely devoted to the theory of absorption, luminescence, and scattering of light in activated crystals. His dissertation was devoted to the theory of vibrational structure in impurity spectra of crystals and the quantum-mechanical method of moments in spectrum problems. He has been a Corresponding Member of the Academy of Sciences of the Estonian SSR since 1961, and an Academician since 1967.

Since 1968 Rebane has been vice president of the Academy of Sciences of the Estonian SSR. During the preceding four years he was academic secretary of the division of physicotechnical and mathematical sciences of the Academy, and a member of its presidium. At the Institute of Physics and Astronomy he has led the solid state group since 1956 and also the crystal spectroscopy group since 1963. Since 1955 he has taught at Tartu State University, where he has given a number of courses on theoretical physics and solid state physics.

The original Russian text, published by Nauka Press in Moscow in 1968 for the Institute of Physics and Astronomy of the Academy of Sciences of the Estonian SSR, has been revised and corrected for this edition by the author. The translation is published under an agreement with Mezhdunarodnaya Kniga, the Soviet book export agency.

Ребане Карл Карлович

Элементарная теория колебательной структуры спектров примесных центров кристаллов

ELEMENTARNAYA TEORIYA KOLEBATEL'NOI STRUKTURY SPEKTROV PRIMESNYKH TSENTROV KRISTALLOV

Library of Congress Catalog Card Number 69-12540

ISBN-13: 978-1-4684-1778-4   e-ISBN-13: 978-1-4684-1776-0
DOI: 10.1007/978-1-4684-1776-0

© 1970 Plenum Press, New York
Softcover reprint of the hardcover 1st edition 1970

A Division of Plenum Publishing Corporation
227 West 17th Street, New York, N. Y. 10011

United Kingdom edition published by Plenum Press, London
A Division of Plenum Publishing Company, Ltd.
Donington House, 30 Norfolk Street, London W.C.2, England

Preface to the American Edition

It is very rewarding for an author to know that his book is to be translated into another language and become available to a new circle of readers.

The study of the optics and spectroscopy of activated crystals has continued to grow. The development and first remarkable successes of light scattering by impurities in crystals have occurred in the comparatively short time since my original book was sent to press. After experimental observation of the sidebands (wings) in impurity infrared absorption spectra, interest in these spectra as a source of information on the vibrations of a crystal in the neighborhood of an impurity has increased significantly.

Therefore, in addition to making minor corrections, I have supplemented the section on the effect of anharmonicity (section 25) and written two new sections and another Appendix on infrared absorption, scattering of light by an impurity center in a crystal, and the adiabatic approximation, respectively.

The bibliography has received several dozen new entries, but it nevertheless does not pretend to be complete.

I hope that the American edition is useful and in some degree corresponds to the general deepening of our physical understanding of solids.

K. Rebane

Preface to the Russian Edition

Interactions with lattice vibrations are an important problem in contemporary solid-state theory and crystal spectroscopy. These interactions are clearly seen in the vibrational structure of the absorption and luminescence spectra of activated crystals which are, in essence, always vibronic spectra. The properties which determine whether a crystal can be used in a laser depend largely on them.

The vibronic transition problem first arose in its present formulation in the theory of molecular spectra. It is comparatively simple for diatomic molecules. Therefore it was here, on the basis of the Franck—Condon principle, that a correct understanding of the physical situation was achieved, the principle was given a quantum-mechanical basis, and the first quantum-mechanical calculations of the distribution of intensity over the vibrational components of the electronic transition were made [1-4].

The interaction between an optical transition and the vibrations of a polyatomic molecule is quite similar to the problem of an impurity center in a crystal. The need to explain the nature of broad structureless spectral bands whose shapes could be represented by a certain more-or-less universal function applicable to a large number of complex molecules first arose long ago in the theory of the luminescence of complex molecules (Vavilov [5]). The current status of the theory of the shapes of vibronic absorption and luminescence bands in complex molecules, whose development has received major contributions from the work of Neporent [6] and especially of Stepanov and his co-workers, has been developed in monographs by Stepanov [7], and Stepanov and Gribkovskii [8].

The present quantum-mechanical theory of the shapes of absorption and luminescence bands arising from impurities in crystals originated with the work of Pekar [9, 10], Kun Huang and Rhys [11], Davydov [12], and Krivoglaz and Pekar [13]. In these papers the nature of the broad vibronic bands was elucidated, proceeding from the idea, first put forth by Frenkel' [14], that changes of the normal modes with a change of the electronic state could be the reason for the appearance of multiphonon transitions. Effective methods of calculation were developed and formulas were obtained which give a good description of such basic properties of these bands as the Stokes shift, the temperature dependence of the half-width, and Levshin's mirror symmetry law [15, 16].

In connection with the main objective of this book, the theory of spectra with clearly-defined vibrational structure, we take particular note of the work of Krivoglaz and Pekar [13], in which a theory was developed which takes account of the dispersion of the normal mode frequencies in the crystal. In the light of our present understanding, it is especially noteworthy that they showed that a narrow resonant purely electronic line (ideally with the radiation width) exists even with dispersion of the vibration frequencies.* In essence this line is the same kind of analog and "precursor" of the Mössbauer line as the zero-phonon Lamb line in the spectrum of neutrons absorbed by nuclei in a crystal [17].

It cannot be said that the impurity spectra of crystals have been neglected by experimenters. Since the discovery of line spectra in organic crystals by Obreimov [18] and in uranyl crystals by Becquerel [19], many investigations have been made of the absorption and luminescence spectra of a wide variety of inorganic and organic crystals. In our opinion there are three series of papers which are the most interesting in connection with vibrational structure: the spectra of rare-earth ions in ionic crystals [20, 21], investigations of molecular crystals with impurity molecules [22], and investigations of uranyl ion spectra [20, 23].

In recent years interest in the vibronic spectra of crystals has increased substantially. This is chiefly due to the development of crystal lasers.

---

* Unfortunately, Pekar and Krivoglaz did not pay due attention to this line; in essence they only remarked in [13] that there is a transformation of one of the "partial spectra" into a line which corresponds to the purely electronic line.

In a laser which uses a crystal containing impurities as the active medium, the level scheme is the scheme of vibronic levels. A large class of problems which have previously been thought of as more-or-less "purely academic" details of the theory of luminescence have turned out to be part of such realistic areas of study as laser theory. The theory of the shape of a spectral line on which a crystal laser works is the theory of the shape of a vibronic line, in particular a purely electronic line.

The most attention has been given to spectra with a sharp purely electronic line, but the vibrational replicas of the purely electronic line and the broad structureless band are of great interest also. It suffices to mention the development of a laser which operates on the vibrational replicas [24] and the theoretical possibility of making a laser using the F band in colored alkali halide crystals [25].

There is a close analogy between the Mössbauer effect and vibronic transitions [26-28]. The purely electronic line has even received a new name — "the optical analog of the Mössbauer line" [29, 30].

The great interest in the Mössbauer effect has led to a substantial revival of theoretical studies on the role of lattice vibrations in this interesting phenomenon. In a short time significant results were achieved on local lattice dynamics in the vicinity of an impurity atom [31-37] and the effect of various subtle effects of the interaction with vibrations on the properties of the Mössbauer line [37-39]. In the theory of luminescence it has long been understood that the local lattice dynamics about an impurity center play an important role in determining the structure of the spectrum. In recent years the development of theories for both phenomena has proceeded in parallel, and often even simultaneously. A number of investigations have developed a unified theory of zerophonon lines (Mössbauer lines and purely electronic lines) and their vibrational backgrounds [38-41].

The mathematical methods developed for vibronic transitions and many formulas obtained for them are almost directly applicable to the theory of the Mössbauer effect. In essence, the analogy of these two phenomena gives a deeper understanding of the physics of interactions between the vibrations and an impurity center.

In connection with lasers and the Mössbauer effect, vibronic transitions in crystals containing impurities have become of interest for real problems. In our view, this is revealed by rapid growth of interest on the part of experimenters. This has been shown by the large number of publications and the organization of conferences especially devoted to the spectra of crystals which serve as active media for lasers and which have richly detailed vibrational structure in their spectra [41]. Of the experimental investigations already specially set up to make a detailed check of the conclusion that the purely electronic line is an optical analog of the Mössbauer line and should basically have its characteristic properties, we note the work of Gross, Permogorov, and Razbirin [29, 30], and that of Hopfield [42], Fitchen, Silsbee, Fulton, and Wolff [43], and Fitchen et al. [44-47].

* * *

This book is an attempt to give an elementary treatment of the present concepts and basic results of the theory of vibrational structure in the vibronic absorption and luminescense spectra of impurity centers in crystals. In this case "elementary treatment" does not mean that we use only classical and the simplest semiclassical ideas about the physics of the process, or that higher mathematics is not used. On the contrary, the basic conclusions of the theory are derived from a proper quantum-mechanical treatment, since all of the features of interest in the vibrational structure of the spectrum are consequences of quantum effects of the lattice vibrations. These effects cannot be understood on the basis of classical ideas; they are tractable semiclassically, but, as usual, good results can only be had from a quantum-mechanical treatment.

The author believes that "elementary" means a quite thorough development of both the basic quantum-mechanical problem and its solution. It is not necessary for this purpose that the reader's knowledge go beyond elementary ideas of quantum mechanics. Therefore the book is easily accessible to readers who have had elementary quantum mechanics in the usual experimental physics or chemistry curricula. The author also feels that the book may be of use to well-prepared readers who are interested in the theory of crystal spectra.

The book includes several dozen problems and exercises of varying degrees of difficulty.

The author thanks O. I. Sil'd for his great assistance in preparing the manuscript for the press, and I. G. Virko for work on the final editing. The author sincerely thanks V. V. Khizhnyakov, G. S. Zavt, N. N. Kristofel', R. A. Preem, A. P. Purga, and L. A. Rebane for valuable remarks.

Contents

Chapter 1

The Adiabatic Approximation

§ 1. The Adiabatic Approximation [3, 4, 48]

The very simplest molecule — the ionized hydrogen molecule H_2^+ — is a complicated system. Even if we consider only the Coulomb interaction we have a three-body problem. An exact solution for such problems is lacking both in classical and quantum mechanics. Therefore we proceed using reasonable approximations in the theory of molecules and crystals. The theory should be quantum-mechanical, but the usual quantum-mechanical perturbation method is not fully appropriate since the expression for the potential energy of a molecule does not contain a small term which can be neglected to allow solution of the Schrödinger equation exactly, so that it could then be taken into account in the first and second orders of perturbation theory.

Therefore we use the a d i a b a t i c approximation as a basis for the quantum-mechanical theory of molecules and crystals. It is based on the substantial difference between the electron mass m and the nuclear mass M. For hydrogen atoms in a crystal or molecule we have M/m = 1840, and for heavier atoms this ratio is still larger.

Using, for the moment, an intuitive classical picture of intramolecular motion, we will speak of the nuclear and electronic velocities in the usual sense. We see intuitively that the heavy nuclei move much slower than the electrons, which are thousands

1

of times lighter. During a time Δt in which a nucleus moves a small distance ΔR, the electrons pass around their orbits many times. Therefore the following approximation suggests itself: we first find the electron orbits for all possible configurations of the stationary nuclei (these orbits will depend parametrically on the nuclear coordinates) and then the motion of the nuclei in a certain average (independent of the electron coordinates) field of the electrons.

We will not give a quantum-mechanical derivation of the adiabatic approximation, but we note that we can arrive at the adiabatic approximation using quantum-mechanical perturbation theory with $(m/M)^{1/4}$ as the expansion parameter. One must carry the approximation to fourth order (see, for example, [48]). Such a treatment reveals the connection between the adiabatic approximation and perturbation theory and shows that in a number of problems the perturbation theory approach is actually inappropriate.

We will now find the Schrödinger equation for a crystal (molecule) (see, for example, [49, 50]). We will use the usual method.

To begin with we use the classical model. The classical model of a crystal is a system of n electrons and N nuclei which interact by Coulomb forces. We denote the electron mass by m, the mass of the α-th nucleus by M_α, the electron charge by $-e$, and the α-th nuclear charge by $Z_\alpha e$.

We now write the classical hamiltonian for such a system. The hamiltonian H for a closed system in which there are no dissipative forces is the total energy of the system expressed in terms of the coordinates and their conjugate momenta. The energy $\mathscr{E}$ of the system is the sum of the kinetic energies of motion of all the electrons and nuclei and their potential energy of interaction

$$\mathscr{E} = \frac{m}{2} \sum_{i=1}^{n} \mathbf{v}_i^2 + \frac{1}{2} \sum_{\alpha=1}^{N} M_\alpha v_\alpha^2 + V(r, R),$$

where r is the set of all electron coordinates, and R is the set of all nuclear coordinates. The potential energy $V(r, R)$ of the interaction of the electrons and the nuclei is given by

$$V(r,R) \equiv \frac{1}{2} \sum_i \sum_j \frac{e^2}{|r_i - r_j|} + \frac{1}{2} \sum_\alpha \sum_\beta \frac{e^2 Z_\alpha Z_\beta}{|R_\alpha - R_\beta|} - \sum_\alpha \sum_i \frac{e^2 Z_\alpha}{|R_\alpha - r_i|}.$$

Here the sum runs over the coordinates of all electrons and nuclei except where $i = j$ or $\alpha = \beta$.

In cartesian electron coordinates x_i, y_i, z_i the conjugate momenta are $p_{ix} = mv_{ix}$, $p_{iy} = mv_{iy}$, and $p_{iz} = mv_{iz}$; similarly for the α-th nucleus $P_{\alpha x} = M_\alpha v_{\alpha x}$, $P_{\alpha y} = M_\alpha v_{\alpha y}$, and $P_{\alpha z} = M_\alpha v_{\alpha z}$. Consequently the hamiltonian for our system is

$$H = \frac{1}{2m} \sum_{i=1}^n p_i^2 + \frac{1}{2} \sum_{\alpha=1}^N \frac{P_\alpha^2}{M_\alpha} + V(r, R).$$

We will go from the function H to the hamiltonian operator $\hat{H}$. Since in quantum mechanics the coordinate x corresponds to multiplication by x, and the momentum p_x corresponds to the operator $\hat{p}_x = ih\partial/\partial x$,* the operator $\hat{H}$ is written as

$$\hat{H} = -\frac{h^2}{2m} \sum_{i=1}^n \Delta_i - \frac{h^2}{2} \sum_{\alpha=1}^N \frac{\Delta_\alpha}{M_\alpha} + V(r, R), \qquad (1.1)$$

where

$$\Delta_i = \frac{\partial^2}{\partial x_i^2} + \frac{\partial^2}{\partial y_i^2} + \frac{\partial^2}{\partial z_i^2}$$

is the Laplacian operator.

The Schrödinger equation for the stationary states of the system, as we know from quantum mechanics, can be constructed as

$$\hat{H}\psi(r, R) = E\psi(r, R),$$

where ψ is the wave function and E is the energy eigenvalue (energy level).

The problem thus reduces to solving the differential equation

$$-\frac{h^2}{2m} \sum_i \Delta_i \psi - \frac{h^2}{2} \sum_\alpha \frac{\Delta_\alpha \psi}{M_\alpha} + V(r, R)\psi = E\psi. \qquad (1.2)$$

---

* Throughout the book h corresponds to Planck's constant divided by 2π, usually denoted by ħ in English texts — translator.

We will solve Eq. (1.2) in the adiabatic approximation (see, for example, [4, 48, 50]). It should be noted that an exact solution of (1.2), even if we could find one, would be of little use in reaching physical conclusions because of its very complex form.

We will seek a solution of Eq. (1.2) as a product of two functions

$$\psi(r, R) = \Phi(r, R)\, \varphi(R).\tag{1.3}$$

We will see below that the function Φ is an electronic wave function which depends parametrically on the nuclear coordinates, and φ is a wave function which describes the nuclear motion. Substituting (1.3) into (1.2) we find

$$-\frac{\hbar^2}{2m}\varphi(R)\sum_i \Delta_i \Phi(r, R) - \frac{\hbar^2}{2}\sum_\alpha \frac{\Delta_\alpha}{M_\alpha}\Phi(r, R)\varphi(R)$$
$$+ V(r, R)\Phi(r, R)\varphi(R) = E\Phi(r, R)\varphi(R).\tag{1.4}$$

We will also neglect the terms in Eq. (1.4) which contain a differentiation of the electron wave function $\Phi(r, R)$ with respect to the nuclear coordinates, i.e., the terms

$$\hat{L}\psi(r, R) = -\frac{\hbar^2}{2}\sum_\alpha \frac{1}{M_\alpha}[\varphi(R)\,\Delta_\alpha\Phi(r, R) + 2\,\nabla_\alpha\varphi(R)\nabla_\alpha\Phi(r, R)].\tag{1.5}$$

The operator $\hat{L}$ defined by Eq. (1.5) is called the n o n a d i a b a t i - c i t y   o p e r a t o r.

The solution of the Schrödinger equation (1.4) now reduces to the solution of two equations. We can separate each term in this equation into the product of the wave functions, $\Phi(r, R)\varphi(R)$. Then we find an equation in which we have terms containing only nuclear or only electronic wave functions. This equation can be written as a system of two equations

$$-\frac{\hbar^2}{2m}\sum_i \Delta_i \Phi(r, R) + V(r, R)\,\Phi(r, R) = W(R)\,\Phi(r, R),\tag{1.6}$$

$$-\frac{\hbar^2}{2}\sum_\alpha \frac{\Delta_\alpha}{M_\alpha}\varphi(R) + W(R)\varphi(R) = E\varphi(R).\tag{1.7}$$

Each equation has the form of a Schrödinger equation. The first equation of the system describes the stationary states of the elec-

trons in the adiabatic approximation. It is a differential equation
in the electronic coordinates. There is no differentiation with respect to nuclear coordinates; therefore Eq. (1.6) can be regarded
as a set of Schrödinger equations for the electrons in the crystal
with nuclei fixed at positions R. For each nuclear position R we
have an electronic equation of motion. The energy eigenvalue
W(R) of the electronic state (it is evident from our definition of the
operator V(r, R) that W(R) also includes the potential energy of
the interaction between nuclei) depends on the nuclear positions.
We say that the electronic Schrödinger equation (1.6) depends parametrically on the nuclear coordinates. Equation (1.6) is very
complicated and it has only been solved exactly for the H_2^+ molecule (and other ionized diatomic molecules with a single electron).

Equation (1.7) is a Schrödinger equation which governs the
motion of the nuclei in the molecule. The function W(R), the eigenvalue of the electronic equation which depends parametrically on
the nuclear coordinate, plays a large role in this equation. In
Eq. (1.7), which we will also call the vibrational equation, it plays
the role of the potential energy for nuclear motion. Each eigenvalue of Eq. (1.7) is also an eigenvalue of the total energy of the
molecule in the adiabatic approximation.

The degree of complexity of the nuclear equation of motion
is determined by the complexity of W(R). Its exact form is not
known, but in one exceedingly important case — that of small oscillations of the nuclei — we know it approximately. The success
of the adiabatic approximation in the theory of crystals and molecules arises from just this fact, and in particular, as we shall see,
the success of the theory of vibronic spectra also.

Thus the wave function which describes the n-th stationary
state of a system in the adiabatic approximation (i.e., neglecting
the nonadiabaticity terms (1.5)) can be written in the form (1.3)
where Φ and φ are eigenfunctions of the Schrödinger equations (1.6)
and (1.7)

$$\psi_n(r, R) = \Phi_l(r, R)\varphi_{li}(R). \qquad (1.3a)$$

Here l and i enumerate the eigenfunctions and eigenvalues
of the electronic (1.6) and nuclear (vibrational) (1.7) Schrödinger
equations.

We rewrite these equations showing the quantum numbers explicitly

$$-\frac{\hbar^2}{2m}\sum_l \Delta_l \Phi_l(r, R) + V(r, R)\,\Phi_l(r, R) = W_l(R)\Phi_l(r, R); \qquad (1.6a)$$

$$-\frac{\hbar^2}{2}\sum_a \frac{\Delta_a}{M_a}\,\varphi_{li}(R) + W_l(R)\,\varphi_{li}(R) = E_{li}\varphi_{li}(R). \qquad (1.7a)$$

Since the potential energy operator in (1.7a) is an eigenvalue of Eq. (1.6a), it depends on the electronic quantum number l. This means that the vibrational functions φ depend on l, which is reflected in a second index for the functions φ_{li}. The index l does not have all the usual properties of a quantum number. Thus the functions φ_{li} and $\varphi_{l'i}$ are not mutually orthogonal. This is of fundamental significance in the theory of vibronic transitions.

Equation (1.7), which describes the nuclear motion, also contains translational and rotational motion of the molecule (crystal). For crystals, which are of interest here, these motions can be neglected almost completely. An exception is the unusual occurrence of an internal rotational degree of freedom in a crystal.

It has been shown [51] how the effect of the translational and rotational motion on the vibronic spectrum decreases with increasing molecular mass.

We will assume that the problems of translational and rotational motion have been solved, the results taken into account, and that by a suitable choice of coordinate systems they have been eliminated from further consideration. Equation (1.7), which now describes the vibrational motion of the nuclei, will be called the nuclear vibration equation.

A more careful theoretical treatment allows us to improve Eq. (1.7) somewhat [48]. The improvement means that for the potential energy operator, instead of $W_l(R)$ we use a $U_l(R)$, which is given by

$$U_l(R) = W_l(R) - \frac{\hbar^2}{2}\sum_a \frac{1}{M_a}\int \Phi_l(r, R)\,\Delta_a\Phi_l(r, R)\,dr. \qquad (1.8)$$

The second term in the improved potential energy operator which describes the nuclear motion in the adiabatic approximation represents the nuclear kinetic energy operator averaged over the electronic wave function. This term arises from the (nonlinear) dependence of the electronic wave function on the nuclear coordinates. We assume that it is a small correction to the operator $W_l(R)$.

In view of the importance of the adiabatic approximation in the present theory of molecules and crystals, we have given a derivation of the basic equations from a variational principle in Appendix VII, for readers interested in the principles underlying the adiabatic approximation. This derivation gives the adiabatic approximation a precise form, i.e., the potential energy operator for the heavy particles, $U_l(R)$, is in the form (1.8). We further investigate some of the properties of the usual (imprecise) adiabatic approximation using the variational principle. We show that the lowest exact energy level E_0 lies between the approximate values calculated using the precise and the usual adiabatic approximations, i.e.,

$$E'_{00} \geqslant E_0 \geqslant E_{00}, \tag{1.9}$$

where E_{00} is calculated from Eqs. (1.6a) and (1.7a), and E'_{00} from (1.6a) and (1.7a) in which $W_l(R)$ has been replaced by $U_l(R)$ from (1.8). This is also known as the Hellmann–Feynman theorem.

Exercise 1.   Illustrate the difference between the electronic and nuclear velocities in a classical picture, by discussing (using classical mechanics):

(a) the velocity of motion (rotation) of the electron and the nucleus about the center of mass of a hydrogen atom,

(b) the vibration of harmonic oscillators having given force constants but different masses,

(c) motion of free particles having different masses under thermal equilibrium conditions, taking into account the classical equipartition of kinetic energy among degrees of freedom.

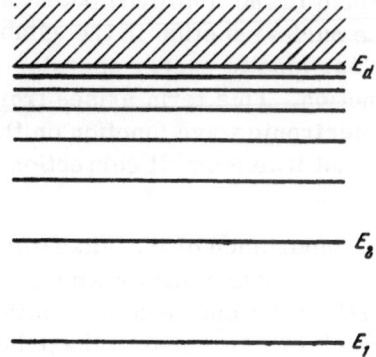

Fig. 1. Schematic molecular energy
levels.

§ 2.   Molecular Energy Levels and Potential Curve Diagrams. The Harmonic Approximation

We will repeatedly turn to potential curve diagrams for
clarity. Near an impurity center the strict periodicity in the spatial positions of the particles in the crystal is violated, and translational symmetry, which plays a paramount role in processes relating to excitation of the host material, recedes into the background for processes localized at an impurity center. Therefore
an impurity center (i.e., a structure consisting of an impurity
atom, ion, or molecule, plus a certain number of lattice particles
in the vicinity of the impurity) has many of the distinguishing features of a molecule. In considering these properties of an impurity center, the potential curve diagram turns out to be a useful aid
to thought. Therefore it is appropriate to recall some features of
the potential curves of diatomic molecules.

From the quantum-mechanical point of view, the energy levels E_i of a molecule are the eigenvalues of the total energy operator $\hat{H}$ (1.1) of the molecule. In order to find the E_i in the adiabatic
approximation, we must first find the eigenvalues $W(R)$ of the electron energy operator and then solve Eq. (1.7). The eigenvalues
obtained will be the total energies of the molecule in the adiabatic
approximation. In Fig. 1 we show schematically the energy levels
of a molecule. The place where the levels merge into the contin-

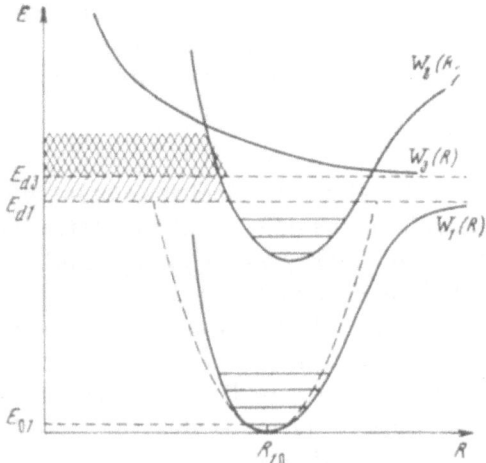

Fig. 2. The potential curve diagram for a diatomic molecule. It can also be conditionally applied to localized or pseudolocalized vibrations of an impurity center.

uous spectrum corresponds to ionization or dissociation of the molecule; E_d is the least energy required to separate any part of the molecule to infinity. The scheme of Fig. 1 is for diatomic molecules. Polyatomic molecules (and especially crystals) have many vibrational degrees of freedom and correspondingly many energy levels which are particularly dense at high energies. Some parts of the energy spectrum become quasicontinuous.

Experiments give information on the transition probabilities and not directly on the energy levels. The potential curve and the potential curve diagram are very useful conceptual tools in the theory of transitions, and especially in the theory of vibronic transitions (i.e., transitions where both the electronic and vibrational states change simultaneously). We will consider these ideas for diatomic molecules.

In contrast to an energy level diagram, which is one–dimensional, the potential curve diagram is two-dimensional; we plot the energy E and the distance R between the nuclei of the diatomic molecule along the axes (Fig. 2). The potential curve $W_l(R)$ is the l-th eigenvalue of the electronic equation (1.6) which depends parametrically on the vibrational coordinate R. In the Schrödinger equation (1.7) for the nuclei, $W_l(R)$ acts as the potential energy operator for nuclear motion. In classical and semiclassical treatments (as we will see below, they are fruitful in the theory of vibronic transitions) $W_l(R)$ is simply the potential energy curve for the relative motion of the nuclei.

The curve $W_i(R)$ shows an electronic state in which a stable state of the molecule is possible. In the classical approximation E_{d1} is the dissociation energy and R_{10} is the equilibrium distance between the nuclei. In a quantum treatment the dissociation energy is given by $E_{d1} - E_{01}$, where E_{01} is the zero-point vibration energy; the coordinate R_{10} of the minimum of the potential energy operator is close to the average distance between the nuclei in the electronic ground state if the molecular vibration is weakly excited. $W_2(R)$ is the same kind of potential curve for an excited electronic state. A potential curve without a minimum, such as $W_3(R)$, corresponds to an excited electronic state in which the atoms do not form a molecule.

In the theory of vibronic optical spectra at modest temperatures we are particularly interested in systems in which the molecule is in a state of thermal equilibrium and undergoes small vibrations before the transition. In describing the small vibrations, it suffices to consider the potential curve near the equilibrium position R_{l0}, expanding it in a Taylor series up to and including quadratic terms. (In Fig. 2 the corresponding parabola is shown by a dashed line.)

$$W_l(R) = W_l(R_{l0}) + \frac{1}{2}\left(\frac{d^2W_l}{dR^2}\right)_{R=R_{l0}} (R - R_{l0})^2 + \dots$$
$$\equiv W_l(R_{l0}) + \frac{1}{2} a_l (R_l - R_{l0})^2 + \dots . \qquad (2.1a)$$

The vibration frequency of the molecule in the l-th electronic state is related to the coefficient of the quadratic term in the expansion (2.1a) by $\omega_l = (a_l/M)^{1/2}$ where M is the reduced mass of the molecule. For the electronic ground state of a simple diatomic molecule, a_l and ω_l can be calculated rather precisely [52]. This is also possible for the first excited electronic state, but already with less precision. Another possibility for solving the problem lies in a semi-empirical approach: the constant a_l is regarded as a parameter of the theory and the properties of the spectrum are interpreted on the basis of (2.1a); from a comparison of these properties with the measured spectrum we find numerical values of ω_l and a_l. In a given case ω_l can be found, for example, from the distance between vibrational components in infrared absorption spectra, in vibronic spectra, or in Raman scattering spectra. In diatomic molecules composed of heavy atoms this typical spectroscopic approach becomes preferable. In considering complex polyatomic molecules and impurity centers in crystals it is still a unique approach, which allows one to develop the theory of vibronic spectra with the precision and reliability necessary for present-day experiments. Therefore, we will also use the semi-empirical spectroscopic approach in this book.*

Substitution of (2.1a) into the vibrational equation (1.7) transforms it into the Schrödinger equation of a harmonic oscillator. The solution of this equation is well known (see, for example [49], and also Appendix II). It has energy levels $E_{li} = W_l(R_{l0}) + \hbar\omega_l(i + \frac{1}{2})$, where the quantum number i assumes the values 0, 1, 2, ... The wave functions $\varphi_{li}(R)$ of the stationary states are Hermite polynomials multiplied by gaussian functions which serve as weight functions (see Eq. (II.52)).

As noted above, in the adiabatic approximation E_{li} are also eigenvalues of the total energy of the molecule. They are shown in Fig. 2 by horizontal lines inside the potential wells (in the region where $E_{li} > W_l(R)$). These crudely characterize the spatial extent of the wave functions, which, in the region where $E_{li} < W_l(R)$ (corresponding classically to negative kinetic energy), drop off rapidly beyond the turning point $E_{li} = W_l(R)$.

We will now turn to polyatomic molecules and crystals. In more complicated molecules, having N vibrational degrees of freedom (N = 3n — 6, where n is the number of atoms in the molecule), $W_l(R)$ is a function of N coordinates. Instead of potential curves we have N-dimensional potential surfaces and the analog of Fig. 2 is an (N + 1)-dimensional "diagram" with coordinates E, R_1, R_2, ... R_N. The physical meaning of $W_l(R)$ as the eigenvalue of the electronic equation (1.6) and the potential energy operator in

---

* See [53-55] for the present status of theories in which the fundamental characteristics of $W_l(R)$ for impurity centers are calculated using only the fundamental constants of physics.

Eq. (1.7) remains as before. To emphasize the connection between the ideas of "potential curve" and "potential surface" and the adiabatic approximation, we will always use the term "adiabatic potential".

If a large molecule (crystal) is in a stable state then the vibrational motions of its atoms, as for a diatomic molecule, reduce to small* vibrations about the equilibrium positions and we can restrict our attention to the region of the potential surface near the equilibrium configuration of the molecule.

In this region the function $W_l(R)$ can be expanded in a series in many variables, analogous to (2.1a)

$$
\begin{aligned}
W_l(R) = W_l(R_{l0}) + \frac{1}{2}\Big\{ & \frac{\partial^2 W_l}{\partial R_{1x}^2}(R_{1x} - R_{1x,l0})^2 \\
+ \frac{\partial^2 W_l}{\partial R_{1y}^2}(R_{1y} - R_{1y,l0})^2 & + \frac{\partial^2 W_l}{\partial R_{1z}^2}(R_{1z} - R_{1z,l0})^2 \\
+ \frac{\partial^2 W_l}{\partial R_{2x}^2}(R_{2x} - R_{2x,l0})^2 & + \ldots + 2\frac{\partial^2 W_l}{\partial R_{1x}\partial R_{1y}}(R_{1x} - R_{1x,l0}) \\
\times (R_{1y} - R_{1y,l0}) & + \ldots \Big\}.
\end{aligned}
\tag{2.1b}
$$

Here $R_{\alpha x}$, $R_{\alpha y}$, and $R_{\alpha z}$ are the cartesian coordinates of the α-th atom in the center-of-mass system of the molecule,† and the derivatives are taken at the equilibrium configuration.

After introducing appropriate new variables q_{ls} (normal coordinates) the expression for $W(R)$ simplifies substantially; it takes the form of a diagonal quadratic form, i.e., in this coordinate system the cross terms vanish (see Appendix II). When we write the variables in Eq. (1.7) in normal coordinates, they are separated, and the problem reduces to finding the energy levels and wave functions for the stationary states of a set of N independent harmonic oscillators. We find for the total vibrational energy

$$
E_{l; i_1, i_2, \ldots, i_N} = \sum_{s=1}^{N} \hbar\omega_{ls}\left(i_s + \frac{1}{2}\right) =
$$

---

* Strictly speaking, the amplitudes of the changes in relative position between the neighboring atoms are small in comparison to the corresponding equilibrium separations.

† These coordinates also describe molecular rotation. See [56] for the separation of the rotational motion of the molecule.

$$= \hbar\omega_{l_1}\left(i_1 + \frac{1}{2}\right) + \hbar\omega_{l_2}\left(i_2 + \frac{1}{2}\right) + \ldots + \hbar\omega_{lN}\left(i_N + \frac{1}{2}\right), \quad (2.2)$$

where the index l corresponds to the electronic state, the index $s = 1, 2, \ldots N$ corresponds to the frequency ω_{ls}, and the numbers i_s are the vibrational quantum numbers. The shape of the adiabatic potential changes with a change of the electronic state. Therefore each electronic state will have, strictly speaking, its own system of normal coordinates. In general, there are differences not only in the numerical values of the frequencies ω_{ls}, but also in the forms of the corresponding normal modes (in a different electronic state they are given by different linear combinations of the cartesian coordinates of the nuclear displacements).

The wave function is a product of wave functions for the individual normal oscillators

$$\varphi_{l;i_1,i_2,\ldots,i_N} = \prod_{s=1}^{N} \varphi_{l;i_s}(q_{ls}) = \varphi_{l;i_1}(q_{l1})\,\varphi_{l;i_2}(q_{l2})\ldots\varphi_{l;i_N}(q_{lN}). \quad (2.3)$$

Because the normal oscillators are independent we can introduce a potential curve (parabola) for each of them and give a two-dimensional potential curve diagram. However, it should be kept in mind that, in general, to describe a potential surface we require the system of normal coordinates. Therefore it does not always make sense to speak of the potential curve for a given normal mode in d i f f e r e n t electronic states; the system of normal coordinates itself changes with a change of the electronic state. This fact (sometimes called scrambling of the normal coordinates in an electronic transition) is of primary importance in many aspects of the vibronic transitions of molecules and impurity centers in crystals.

What we have said about molecular vibrations also applies to crystals. However, the vibrations of a crystal containing an impurity center have certain additional important properties which follow from the presence of n particles in the crystal, where n is a huge number, of the order of Avogadro's number. These properties, which are treated in the next section, determine the general nature of the vibrational structure in the spectrum of a crystal containing impurities.

The periodicity of the crystal structure is also important, but the basic conclusions remain valid even without strict periodicity. Therefore many of the results obtained for crystals containing impurities are also valid for large molecules if the number of atoms in them is sufficiently large [51].

§ 3. Lattice Vibration Theory for a Crystal Containing Impurities. Band and Localized Vibrations. Pseudolocalized Vibrations

In this section we will give, following [13], a brief discussion of the normal modes of a crystal which contains impurities. Some further details will be found in Appendix II.

The eigenfrequency spectrum for the normal modes of an ideal periodic crystal is composed of alternating allowed and forbidden bands. The number of allowed bands is three times the number of atoms per unit cell of the crystal. The number of frequencies in the band equals the number of unit cells in the crystal, so that the total number of frequencies equals the number of vibrational degrees of freedom, $N = 3n$,* where n is the number of atoms in the crystal. Each normal mode of an ideal periodic crystal is a sinusoidal displacement wave of the atoms, characterized by a wave vector K (see, for example, [57]).

The cartesian coordinates of the atomic displacements are linear combinations of the normal coordinates, and conversely, the normal coordinates are linear combinations of the cartesian ones. It follows from the unitarity of the transformation between these coordinate systems that the coefficients of both of these unitary transformations are of the order of $N^{-1/2}$. Therefore, if the atoms in a limited region of a crystal are displaced, corresponding to a finite change of a finite number of cartesian coordinates, then in the limit $N \to \infty$ the change of each normal coordinate is vanishingly small. If we make a finite change of a finite number of normal coordinates, then the changes of all of the cartesian coordinates are infinitesimal.

We now introduce one impurity molecule, which can be composed of many atoms undergoing intramolecular vibrations, into

---

* Strictly speaking, if we consider the three rotational and three translational degrees of freedom, the number is $N = 3n - 6$.

the crystal and consider how the normal modes of the crystal change.

We distinguish two cases:

1. The intramolecular vibration frequency (somewhat altered by the crystalline surroundings) lies in one of the allowed frequency bands of the ideal crystal and is in resonance with certain eigenfrequencies of the crystalline host lattice. In this case the impurity molecule will radiate elastic waves into the crystal, thus losing energy. It is clear that the intramolecular vibration cannot be a normal mode of the system (crystal + impurity molecule) — only the atoms of the impurity molecule enter into it, but there is coupling between the atoms of the impurity molecule and the atoms of the crystal. The normal modes of the system (crystal + impurity molecule) which have frequencies in an allowed vibrational band of the crystal correspond to motion in which all of the atoms in the crystal participate more or less equally. Since the normal mode energy is finite and all N atoms participate in it in about the same way, the energy of each atom is of the order of N^{-1} and the cartesian displacements are of the order of $N^{-\frac{1}{2}}$. This also applies to the atoms of the impurity molecule. All coefficients in the linear transformation relating the cartesian and normal coordinates are infinitesimals of order $N^{-1/2}$, as in the ideal crystal.

2. The intramolecular vibration frequency (also somewhat altered by the effect of the crystalline surroundings) lies in one of the forbidden bands in the frequency spectrum of the ideal host crystal. Such a vibration cannot resonate with the frequencies of the undisturbed regions of the crystal and does not encounter conditions for propagation outside of the impurity molecule. The molecule will not radiate elastic waves into the crystal.

The intramolecular vibration is not a normal mode of the system (crystal + impurity molecule), since, because of the coupling between the molecule and the crystal, any vibrational motion inside the molecule shakes the crystal atoms surrounding it to some degree. However, it is very important that in the nonresonant case not all of the atoms in the crystal participate in the normal vibration which represents an "intramolecular vibration plus its continuation into the crystal," but only those located near the molecule, since the vibration amplitudes of the atoms in the crystal drop off rapidly with distance between the impurity molecule and the atom. If such a normal mode has a finite energy, the cartesian displacements of the impurity and its nearest neighbor atoms do not depend on N and are finite when $N \rightarrow \infty$. The coefficients in the linear transformation between the cartesian coordinates and this normal coordinate do not depend on N. Vibrations of this kind are l o c a l i z e d vibrations.

There can be L localized vibrations about an impurity center in a crystal, where L is a small integer. The remaining $N - L$ vibrations of the system (crystal + impurity) are of the kind first considered. We will call them b a n d   v i b r a t i o n s.

The c h a n g e s in the equilibrium positions (and the frequencies and "axes") of the normal modes during an electronic transition play a substantial part in the theory of vibronic bands. These arise from finite changes in the equilibrium positions of the atoms and the interatomic force constants between the atoms in the immediate vicinity of the impurity when its electronic state changes. For localized normal modes neither the shift in the equilibrium position, nor the change in frequency, nor the parameters of the "rotation of the axis" depend on N, and in this sense they are finite. For the band vibrations the shifts in the equilibrium positions are of the order of $N^{-1/2}$, and the shifts in frequency and the rotation of the axes are of the order of N^{-1}, as we can easily show, keeping in mind that these quantities are transformation coefficients of the bilinear terms of the potential energy.

An impurity molecule was considered above for clarity. Localized vibrations, as we know, can arise from any highly localized change of the vibrational properties of the lattice [58-60]; in particular they can arise from the introduction of impurities which have no internal vibrational degrees of freedom — impurity ions or atoms. This case is important for inorganic crystalline phosphors and activated crystals used as laser media.

In interpreting the vibrational structure in the spectrum of a crystal containing impurities it turns out to be useful to distinguish p s e u d o l o c a l i z e d (or quasilocalized) vibrations also. We can give a simple idea of a pseudolocalized vibration as a special situation in the first case above. In a pseudolocalized vibration an intramolecular frequency is in resonance with band vibrations, but the coupling of this vibration with the band vibrations turns out to be very weak. This can happen if: (a) the forces connecting the molecule to the crystal are weak, (b) the character (symmetry) of the intramolecular vibration is unsuitable in the given range of band vibrations, (c) the density of frequencies is low in the region of the phonon spectrum in which the frequency of the intramolecular vibration lies. In these cases transfer of energy from the intramolecular vibration to the crystal is inhibited and despite the resonance of the frequencies it will occur slowly. From the point of view of stationary vibrational states of the system (crystal + impurity molecule), it is not a localized vibration. On the other hand, it is evident from physical considerations that in processes which take place in impurity molecules (for example, in an electronic transition) intramolecular vibrations can be excited which, because of the weakness of the coupling with the band vibrations, will behave almost as true localized vibrations.

The ratio of the time τ for transferring energy to the lattice vibrations to the period T of the intramolecular vibration serves as a criterion. If $\tau/T \gg 1$ then the intramolecular vibration, despite its resonance with lattice vibrations, will have the characteristic properties of a localized vibration. In particular, its manifestation in spectra (vibronic, infrared, etc.) will be quite similar to that of localized vibrations. At the same time it is not a localized vibration of the system (crystal + impurity molecule). Thus we call it a "pseudolocalized vibration."

For impurities having no internal degrees of freedom, the problem of pseudolocalized vibrations is more complicated and is of considerable independent interest. In recent years a number of investigations of pseudolocalized vibrations have been carried out. In the first place we note [31], in which they were studied in connection with the Mössbauer effect (see also [32–36, 61, 62]). The concept of a pseudolocalized vibration has been widely used for interpretation of experimental luminescence and Raman scattering spectra (see the survey [40]). Here one speaks of impurities without internal vibrational degrees of freedom. In this case the pseudolocalized vibration can be a marked distortion of the nature of the band vibrations in the vicinity of the impurity, which is not strong enough to produce localized vibrations.

The appearance of localized vibrations requires that rather rigid conditions be fulfilled. For example, for localized vibrations to appear above the highest phonon frequency of the crystal, the impurity must be substantially lighter (by a factor of 1.5 or 2) than the host atoms or there must be a still larger (by a factor of 3 to 5) increase of the interatomic force constants (see [60]). A lesser perturbation could lead to a localized vibration if there is a gap in the phonon spectrum between allowed frequency bands, but this is fairly uncommon for simple crystalline hosts. For example, according to the calculations of [63] only some of the alkali halide crystals have forbidden gaps in their spectra. Therefore localized vibrations about an impurity center having no internal vibrational degrees of freedom in a simple (i.e., having one or two atoms per unit cell) host rarely arise, although the distortion of the vibrations in the region of the impurity is often substantial.

In the absence of localized vibrations the impurity has only an infinitesimal effect on the vibrational f r e q u e n c y spectrum; the change of each frequency is of the order of N^{-1}. It can be very important, however, that the f o r m s of the band vibrations are distorted around the impurity. Indeed, this determines the local lattice dynamics at the impurity center and the nature of the vibrational structure in the spectrum of a crystal containing impurities. The lack of knowledge of the crystal's frequency spectrum recedes into the background. The papers on pseudolocalized vibrations mentioned above specifically discuss the distortions in the form of the band vibrations by an impurity (or other defect).

It turns out that the distortion of a band vibration about a defect can also be quite substantial when no localized vibration occurs. This means that in the formula relating the normal coordinates q_i to the cartesian coordinates x_i

$$x_i = m_i^{-1/2} \sum_j e_{ij} q_j \qquad (3.1)$$

(m_i are the particle masses; for details see Appendix II) the coefficients e_{ij} differ substantially from their values in the undistorted crystal.* In general, with increasing interatomic force constants or with decreasing impurity mass relative to the mass of the host

---

* Both the moduli and phases of the e_{ij} change substantially about the defect, while only the phase changes far from the defect. The defect's effect can also be interpreted as scattering of band vibrations by the defect. Such an interpretation is useful, in particular, in treating the effect of defects on the lattice thermal conductivity [64, 65].

atom which it replaces, there is a tendency toward an increase of the vibration frequency of the impurity.*

If the change is very sharp, a new frequency appears above the spectrum of frequencies allowed in the undistorted crystal. This frequency corresponds to a new normal mode — a localized vibration. The appearance of a localized vibration is accompanied by important changes in the band vibrations; at atoms near the impurity the coefficients e_{ij} decrease sharply for the band vibrations (see Appendix II.7). The band vibrations are "repelled" from the region of the localized vibration.

If the interatomic force constants or the mass change sharply (but not enough for a localized vibration to appear), then this tendency will be pronounced — the displacements of atoms in the neighborhood of the impurity are described by a narrow wave packet of normal modes associated with a pseudolocalized vibration.

For example, for a heavy substitutional isotopic impurity (i.e., an impurity whose introduction does not change the force constants and affects the lattice vibrations only because of the mass change) the frequency $\tilde{\omega}$ and the width Γ of the pseudolocalized vibration "level" are given by the simple formulas [31, 32]:

$$\tilde{\omega} \approx \frac{\omega_D}{\sqrt{3m'/m}} \, , \qquad (3.2a)$$

$$\frac{\Gamma}{\tilde{\omega}} \approx \frac{\pi}{2} \frac{\tilde{\omega}}{\omega_D} \, , \qquad (3.2b)$$

where m and m' are the host and impurity masses respectively. (Equation (3.2) is valid when m' $\gg$ m.) Here ω_D is the Debye frequency, i.e., the highest frequency in the vibration spectrum of the host crystal.[†] Note that the factor which gives the pseudolocalized

---

* The discussion is based on the simple idea that the general tendency is the same as for a harmonic oscillator whose frequency is $\omega = \sqrt{a/m}$, where a is the elastic constant and m is the mass. This has been demonstrated by calculations of the distortion of the band vibrations by an impurity.

† We have in mind a crystal whose unit cell contains one atom. The Debye approximation, which treats the lattice vibrations as waves in a continuous medium, gives a fairly good description of long-wavelength, low-frequency vibrations; pseudolocalized vibrations arising from heavy impurities appear in this region.

vibration a fairly long lifetime in the present case is the low density of frequencies in the low-frequency region of the host crystal spectrum.

The heavy isotopic impurity model is evidently a good approximation to the actual situation for heavy impurity atoms in metals. The corresponding pseudolocalized vibration problem is realistic in the theory of the Mössbauer effect of heavy impurity nuclei in metals (see [32]).

For a number of processes, including vibronic transitions, which involve an impurity center and participation of the vibration of the surrounding lattice particles, linear combinations of the cartesian coordinates of several atoms having definite symmetry are important. Thus the vibrational structure in the spectrum is determined by the way the particular set of cartesian coordinates is expressed in terms of the normal coordinates. In other words, it is determined by the composition of the wave packet of normal modes corresponding to the given set of cartesian displacements. For example, the vibrational structure of the vibronic spectrum in the basic model (see Chapter II) is determined, as we will see below, by the expansion coefficients of the set of cartesian displacements of the equilibrium positions arising from the electronic transition under study, in the normal modes. If there are localized vibrations, the expansion of $F(\omega)$ contains discrete frequencies corresponding to the localized vibrations, in addition to the continuous spectrum. These discrete frequencies appear in the structure of the spectrum. If there are no localized vibrations but there is a sharp change in the lattice dynamics in the vicinity of the impurity, there can be quite sharp peaks in the continuous spectrum. These peaks, which can arise from pseudolocalized vibrations, can appear in the vibrational structure of the spectrum as quite sharp maxima, similar to the contributions from localized vibrations.

A somewhat more detailed discussion of pseudolocalized vibrations appears in Appendix II, section 8. A summary of the vibrations of a crystal containing impurities and their changes during a change in the electronic state of the impurity, which we need in the theory of vibronic spectra, is given in Table 1.

With this we conclude our brief discussion of the properties of the stationary vibrational states of a crystal and the nature of

TABLE 1. Summary of the Vibrations of a Crystal Containing Impurities and Their Changes during a Change in the Electronic State of the Impurity

Kind of description of the vibrations	Normal mode of the crystal		Packet of band normal modes
	localized	band	pseudolocalized
	width of the level, $\Delta\omega$		
Harmonic approximation	$\Delta\omega = 0$	$\Delta\omega = 0$	$\Delta\omega$ has a finite width determined by the width of the wave packet. The greater the ratio $\bar{\omega}/\Delta\omega$, where $\bar{\omega}$ is the average frequency in the wave packet, the more pronounced the effects of the pseudolocalized vibration. With $\Delta\omega > \bar{\omega}$ it makes no sense to speak of a pseudolocalized vibration.
Relaxation arising from anharmonic interactions taken into account	$\Delta\omega_{anh}$ is finite; the condition $\Delta\omega_{anh} \ll \omega$, where ω is the localized vibration frequency, is the criterion for a satisfactory description of the vibration in terms of the harmonic approximation.	$\Delta\omega_{anh}$ is finite; the condition $\Delta\omega_{anh} \ll \omega$, where ω is the vibration frequency, is the criterion for a satisfactory description in terms of the harmonic approximation.	$\Delta\omega_{anh}$ is finite. The total width of the level is the sum $\Delta\omega + \Delta\omega_{anh}$; with $\Delta\omega_{anh} \gg \Delta\omega$ the pseudolocalized vibration is equivalent to a localized vibration.

Table 1 (continued)

Localized change of the properties of the vibration arising from an electronic transition of the impurity center	Changes of the normal and pseudolocalized vibration parameters		
Finite shifts of the equilibrium positions of a small number of atoms	Finite shifts of the equilibrium positions	Infinitesimal shifts of the equilibrium positions, of the order of $N^{-1/2}$	Finite shifts of the equilibrium positions of the oscillators associated with pseudolocalized vibrations.
Finite shifts of the interatomic force constants between a small number of atoms.	Finite shifts of the frequencies and the principal axes	Infinitesimal shifts of the frequencies and the principal axes, of the order of N^{-1}	Finite shifts of the frequencies and principal axes of the coordinates effective in the pseudolocalized vibrations

of their changes in an electronic transition. We will not dwell on the properties of the electronic states. We first return to calculating vibronic transition probabilities and explaining how these characteristics of the electronic states determine the vibrational structure of interest to us. We see that we only need certain quantities which are a v e r a g e d over the electronic wave functions. They are conveniently treated as parameters of the theory, to be determined experimentally. Thus we do not need to consider a complicated picture of the multifarious electronic properties of the impurity center, and are able to study the vibrational structure of the spectrum in "pure form." The vibrational motion, as we mentioned above, is described to a good approximation in the language of normal modes as the motion of independent quantum objects — harmonic oscillators. This is why the semiempirical spectroscopic approach to the problem has been successful.

§ 4.   Vibronic   Transition   Probabilities.

The   Franck — Condon   Principle

The structure in a vibronic spectrum is determined by the corresponding vibronic transition probabilities. The vibrational structure in the spectral region corresponding to a given electronic transition is determined by the distribution of the transition probability over the vibrational sublevels of the electronic states under consideration.

A leading role is played by the Franck—Condon principle, which in a sense replaces the usual spectroscopic selection rules in the theory of the vibrational structure of vibronic spectra of crystals and molecules. We begin with a brief account of the classical and semiclassical versions of the principle, which are very useful for qualitative interpretations of vibronic processes, in conjunction with potential curve diagrams. Then we give a detailed development of the fundamentals of the quantum-mechanical calculation of the vibrational structure. Appendix I contains certain further details which illustrate the accuracy of the semiclassical version of the Franck—Condon principle, along with the quantum-mechanical results.

a) The Classical Franck—Condon Principle. In its simplest formulation the Franck—Condon principle can be obtained from purely classical considerations. It consists of the assertion that the electronic state (state of the "fast" subsystem) of the molecule (crystal) changes so fast that the nuclei do not move during the electronic transition (condition 1), nor change their momenta (condition 2) ("the slow subsystem does not react"). If the vibronic transition is represented on the potential curve diagram (Fig. 3) by an arrow whose base shows the energy and the coordinate of the vibration before the electronic transition and whose point shows the energy and coordinate at the instant following the electronic transition, then the first condition (R = const.) means that the arrow must be vertical. The second condition (P = const. or $\mathscr{E}_{kin}$ = const.) means that the point of the arrow should lie at the same distance (in energy) from the potential curve of the final electronic state as the base of the arrow from the potential curve of the initial state (in the figure, AA' = BB').

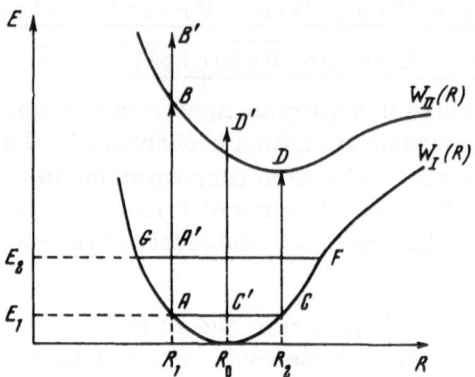

Fig. 3. Potential curve diagram and the Franck—
Condon principle.

To these two conditions, which follow quite naturally from
the basic idea of the adiabatic approximation, we must add another
condition.

According to the classical laws of motion an oscillator
spends most of the time at the turning points (points A and C). We
assume that at the instant of the electronic transition the oscillator is found o n l y at these turning points (condition 3). On a potential curve diagram this condition means that an arrow can only
begin on a potential curve. For the vibrational state with energy
E_1 (Fig. 3) we assume that arrows begin only at points A and C,
and that point C' is "forbidden." For the energy E_2 point A' is
"forbidden" and arrows can begin at points G and F. In agreement
with condition 2, the ends of the arrows should now lie on the potential curves since $\mathscr{E}_{kin} = 0$ before and after the transition (arrows AB and CD).

Thus the classical Franck—Condon principle involves three
statements: (1), R = const.; (2), P = const.; (3), the arrows begin
at the turning points for classical motion of the oscillator.

A simple picture like Fig. 3 already shows the general features of the intensity distribution in the vibronic absorption bands
of diatomic molecules (see [3]).

(b) The Semiclassical Franck—Condon Principle. Parts (1)
and (2) of the classical Franck—Condon principle, as previously

noted, are closely connected with the adiabatic approximation. We might assume (and this is supported by special investigations (see Appendix I)) that this condition gives a good account of the true situation in vibronic transitions. Finally, in quantum mechanics, it is not possible to simultaneously fix the position and momentum of a particle, but one expects that for a sufficiently massive particle, such as a nucleus, quantum effects will not be pronounced.

The weak link in the classical formulation is part (3), which was introduced in addition to the adiabatic approximation, as a rather poor basis for the other requirement.

The semiclassical formulation preserves parts (1) and (2), but replaces part (3) by a new condition: an electronic transition can occur for any value of the vibrational coordinate R with a probability $w(R)$, where $w(R)$ is the quantum-mechanical probability density.* If we are interested in transitions from one particular initial vibrational state, then the probability density is simply the absolute square of the wave function in the coordinate representation, $w_i(R) = |\varphi_i(R)|^2$. If the initial state represents a set of identical systems in thermal equilibrium, we have (for vibrations of a diatomic molecule in the harmonic approximation, or for a single normal mode)

$$w_T(R) = F(T) \sum_i e^{-\frac{E_i}{kT}} |\varphi_i(R)|^2 = \frac{1}{\varepsilon \sqrt{\pi}} e^{-\frac{R^2}{\varepsilon^2}}, \qquad (4.1)$$

where E_i is the energy of the i-th vibrational level,

$$\varepsilon^2 \equiv \frac{h}{m\omega} \coth \frac{h\omega}{2kT},$$

and $F(T)$ is the normalizing factor for the Boltzmann distribution.

The first case is often realized for light diatomic molecules having large vibrational quanta ($h\omega \approx 0.1$ eV). In these molecules even at room temperature ($kT \approx 0.025$ eV) the higher vibrational levels ($i = 1, 2, \ldots$) are hardly excited (90% of the molecules are in the $i = 0$ state). This also holds for high-frequency localized vibrations in crystals. For the band vibrations, which always

---

* There are also other formulations of the semiclassical Franck–Condon principle which are similar to that considered here (see [66, 67]).

include low frequencies, even at liquid helium temperatures we must use the function $w_T(R) = \prod_s w_{sT}(q_s)$, where q_s is the s-th normal coordinate and the function $w_{sT}(q_s)$ is defined by Eq. (4.1) with $\omega = \omega_s$. In the limit $T \to 0$ the function w_T defined by (4.1) goes over to $|\varphi_0(R)|^2$, i.e., to the absolute square of the function which describes the zero-point vibrations of the system.

On the potential curve diagram (Fig. 3) for the vibrational state E_i (E_i should now coincide with one of the oscillator energy levels) the semiclassical principle allows both arrows AB and CD as well as C'D' with the first arrow having the probability $|\varphi_i(R_1)|^2$, the second $|\varphi_i(R_2)|^2$, and the third $|\varphi_i(R_3)|^2$.

If $i = 0$, then the arrows forbidden by the classical Franck–Condon principle have the greatest "weight." This is understandable; for low quantum numbers, and especially for $i = 0$ (zero-point motion of the oscillator!) the classical coordinate distribution differs fundamentally from the real (quantum-mechanical) one.

Naturally doubt arises as to whether the statements R = const. and P = const. do not also lead to significant errors when one deals with transitions beginning in the zeroth vibrational level. A special comparison with the results of a quantum-mechanical calculation carried out by the methods described under (c) below leads to the conclusion that the accuracy of the semiclassical Franck–Condon principle depends basically on the quantum number f of the final vibrational level [66] (see Appendix I). If the region realized in the vibronic transition corresponds to large f ($f \gtrsim 10$ to 20), then the semiclassical Franck–Condon principle gives results very close to the quantum-mechanical results.

In luminescent centers such as $KCl-Tl^+$ the minimum in the potential curve R_{0II} is strongly displaced from the point R_{0I} (the Stokes loss is large) and the most intense part of the vibronic band will correspond to levels with $f \approx 15$ to 50 so that the semiclassical Franck–Condon principle will reproduce the actual intensity distribution well (Fig. 4a). If, however, the potential curve for the final electronic state is little displaced (small Stokes loss) and low vibrational levels f are involved, the semiclassical treatment leads to significant errors, although it is still useful for qualitative interpretation of phenomena (Fig. 4b). The concept of the Stokes loss is dealt with in more detail in section 7.

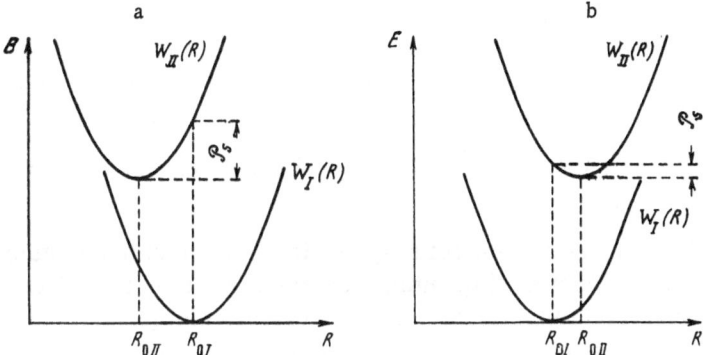

Fig. 4. Potential curves for electronic transitions with large (a) and small (b) Stokes losses P_s.

(c) The Quantum-Mechanical Franck–Condon Principle.

Here we simply calculate the vibronic transition probability according to the rules of quantum mechanics. The results obtained, as we will see at the end of this section, provide a clear comparison with the versions of the Franck–Condon principle given above. Therefore (and, strictly speaking, for this reason only) one speaks of the quantum-mechanical solution as a version of the Franck–Condon principle.

As we know, optical absorption and luminescence processes are quite well described by first-order time-dependent perturbation theory. According to this theory, the transition probability $W_{nn'}$ for a system to go from a state with a quantum number (or set of quantum numbers) n to a state n' is proportional to the absolute square of the corresponding matrix element of the perturbation operator which gives rise to the transition. The perturbation operator $\hat{P}$ will be the dipole moment operator of the molecule. Then $W_{nn'} \sim |P_{n'n}|^2$ and to calculate the transition probability one must construct the matrix element of $\hat{P}$ for the wave functions of the system corresponding to states n and n'.

Below we will only be interested in the function $|P_{n'n}|^2$ as a function of the transition frequency. This function describes both the absorption band and the accompanying emission band. The proportionality coefficient which relates this function to the transition probability (and to the intensity distribution in the spectrum) is much less frequency-dependent than $|P_{n'n}|^2$ and we will not be interested in it. In Appendix V we give a brief treatment of this coefficient, which has a somewhat different form and frequency dependence for absorption and emission.

We will treat the transition probability on the basis of the adiabatic approximation.

We let the state n correspond to an electronic quantum number l and a vibrational quantum number i and the state n' to an electronic quantum number m and a vibrational quantum number f. Then the wave functions of these states are written as [see (1.3a)]

$$\psi_n(r, R) = \Phi_l(r, R)\,\varphi_{li}(R),$$

$$\psi_{n'}(r, R) = \Phi_m(r, R)\,\varphi_{mf}(R).$$

The dipole moment operator $\hat{P}$ for the electronic and nuclear charges takes the form

$$\hat{P} = D_e + D_n = -e\sum_i r_i + e\sum_\alpha Z_\alpha R_\alpha, \qquad (4.2)$$

where r_i and R_α are the multiplicative coordinate operators of the i-th electron and the α-the nucleus. The matrix elements of P take the form

$$P_{n'n} = P_{mf;\,li} = \iint dr\,dR\,\Phi_m(r, R)\,\varphi_{mf}(R)\,[D_e + D_n]$$
$$\times \Phi_l(r, R)\,\varphi_{li}(R) = \int dR\left[\int dr\,\Phi_m(r, R)\,D_e(r)\,\Phi_l(r, R)\right]$$
$$\times \varphi_{mf}(R)\,\varphi_{li}(R) + \int dR\varphi_{mf}(R)\,D_n\varphi_{li}(R)\left[\int dr\,\Phi_m(r, R)\,\Phi_l(r, R)\right].$$

Since Φ_m and Φ_l are essentially two eigenfunctions of the Schrödinger equation (1.6a) which form a system of orthonormal functions, the part of the second term in the second form of the equation which is in square brackets is zero when $l \neq m$ and unity when $l = m$. We know that the dipole moment of the nuclei gives rise only to purely vibrational transitions within a single electronic state. These transitions involve infrared absorption by the vibrations. We will discuss this in section 31. The first term, which contains the dipole moment of the electrons as the perturbation operator, involves vibronic transitions in which the electronic

state changes.* Since the electronic wave functions of the system
are not known, we use the notation

$$D_{ml}(R) \equiv \int dr \Phi_m(r, R) D_e(r) \Phi_l(r, R) \qquad (4.2a)$$

and regard $D_{ml}(R)$ as a parameter of the theory which can be
found from experimental data if needed.

We further consider that the electronic matrix element
$D_{ml}(R)$ is only weakly dependent on the nuclear coordinates and
expand it in a series about the equilibrium position of the nucleus

$$D_{ml}(R) = D_{ml}^0 + D_{ml}^{(1)} R + D_{ml}^{(2)} R^2 + \ldots \qquad (4.2b)$$

If there are many nuclear coordinates then the series in (4.2b) is
a multidimensional Taylor expansion. The C o n d o n a p p r o x i-
m a t i o n consists of replacing the function $D_{ml}(R)$ by the zeroth
term D_{ml}^0 in the expansion, which does not depend on the nuclear
coordinate. The Condon approximation is more precise than the
assumption $\Phi(r, R) = \Phi(r, R_0)$, since by integrating the electronic
wave function over the electronic coordinates we carry out some
averaging, and the dependence on the nuclear coordinate is already weak for $D_{ml}(R)$.

It is not necessary to neglect the subsequent terms in (4.2b);
there are no special mathematical difficulties involved in including
them. If we do include them, then there are new coefficients $D_{ml}^{(r)}$
where $r = 1, 2, \ldots$, which are also regarded as parameters of the
theory to be determined from experiments. We will see later that
the deviation from the Condon approximation is substantial for
vibronic transitions forbidden by symmetry considerations in the
equilibrium configuration of the molecule.

Thus the vibronic transition probability is basically given
(in the Condon approximation) by the following expression

$$W_{li;\,mf} = |D_{ml}^0|^2 \left| \int \varphi_{mf}(R) \varphi_{li}(R) dR \right|^2. \qquad (4.3)$$
$$\scriptstyle l \neq m$$

---

* This term also makes some contribution to the infrared absorption due to purely vibrational transitions. This contribution appears if the integral over electron coordinates with $l = m$ is nonzero, which can occur when the impurity center (molecule)
lacks inversion symmetry.

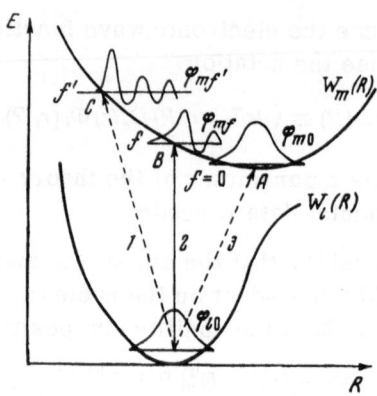

Fig. 5. Relation between the Franck—
Condon principle and the quantum-
mechanical vibronic transition probabilities. The most probable transition is
transition 2, which is allowed by the
Franck—Condon principle; transitions
1 and 3 are substantially less probable
since they correspond to small values of
the overlap integral between the vibra-
tional functions $\varphi_{l0}\varphi_{mf'}$ and $\varphi_{l0}\varphi_{m0}$.

The distribution of the vibronic transition probability over
vibrational states is given by the integral on the right-hand side of
(4.3) which is called the overlap (superposition) integral of the
wave functions.

We again stress that the vibrational functions $\varphi_{mf}(R)$ and
$\varphi_{li}(R)$ are solutions of different Schrödinger equations (1.7)
(the potential energies are not the same, $W_m(R) \neq W_l(R)$) belonging
to different orthonormal systems. Therefore the overlap integrals
in (4.3) do not go to zero when $f \neq i$, if $m \neq l$.

We show that, depending on the overlap integral between i
and f, the classical formulation of the Franck–Condon principle
contained in parts (1) and (2) (R = const., P = const.) appears as
an approximate estimate of the value of the integral. We make a
comparison of the vibronic transition probabilities shown in Fig. 5
by arrows 1, 2, and 3. These transitions are from electronic
state l in a vibrational level i = 0 into an electronic state m with
vibrational quantum numbers f. We consider a case where there

is a substantial shift in the minimum of the upper potential curve. The probabilities of such transitions are proportional to the squares of the overlap integrals of the vibrational wave functions of the initial and final states. It is evident from the figure that the maximum overlap, and consequently the maximum probability, will occur for the transition shown by the vertical arrow 2 (with excitation of f vibrational quanta). The transition which is most probable also agrees with the semiclassical Franck—Condon principle. Transitions 1 and 3 shown by sloping arrows correspond to much lower probabilities. In the language of the semiclassical treatment they correspond to strong violation of the requirements R = const. or P = const.;* in the quantum language they correspond to very small values of the overlap integral. For transition 3 the functions φ_{l0} and φ_{m0} simply do not overlap, and in case 1 the function φ_{l0} overlaps with a strongly oscillating part of the function $\varphi_{mf'}$.

(d) The Franck—Condon Principle as a Selection Rule. We have previously mentioned that the Franck—Condon principle in a sense replaces the selection rules for vibronic transitions. At the same time the Franck—Condon principle is not a selection rule in the usual sense — it does not put any strict limitations on transitions. It merely says that transitions into one vibrational state are highly probable (arrow 2 of Fig. 5), that some are somewhat less probable (into levels whose classical turning points are close to B), and that those into levels with a substantial violation of the Franck—Condon principle are very improbable.

This special feature of the Franck—Condon principle as a selection rule is related to the adiabatic approximation. The use of the adiabatic wave functions in the previous section meant that the matrix element [see Eq. (4.3)]

$$\int \varphi_{mf}(R)\, \varphi_{li}(R)\, dR, \qquad (4.4)$$

which determines the distribution of the probability over vibrational levels, is constructed from vibrational functions which belong to d i f f e r e n t   s e t s of orthonormal eigenfunctions (φ_{li} is an eigen-

---

* We have shown the arrows corresponding to violation of the condition R = const., but one could just as well show vertical arrows corresponding to sharp violation of the condition P = const., or all other possible arrows which correspond to violation of both conditions to one degree or another.

function of one of the Schrödinger equations (1.7a) with a potential energy operator W_l (R) while φ_{mf} is an eigenfunction of another Schrödinger equation with a potential W_m(R).

In quantum-mechanical calculations of transition probabilities we usually deal with a s i n g l e   s e t of orthonormal eigenfunctions. It is just this orthonormality property of the wave functions which enter the matrix elements, in combination with the symmetry properties of the system (the symmetry properties of a s i n g l e hamiltonian), which lies at the basis of the usual spectroscopic selection rules. These selection rules are formulated as "absolute"statements ("the line is there or it isn't") and therefore are also sometimes called exclusion rules.

For example, for dipole transitions of a harmonic oscillator the well-known selection rule states that the vibrational quantum number i can only change by one unit

$$\Delta i = \pm 1. \tag{4.5}$$

All other transitions are forbidden in the dipole approximation.

For vibronic transitions (which relate basically to harmonic oscillators!) this selection rule completely loses its meaning. The Franck–Condon principle which replaces it cannot usually be formulated as an exclusion rule.

However, there is one exception — where the potentials for nuclear motion are the same in both electronic states, i.e., W_l(R) = W_m(R) (no Stokes loss). In this case the sets of vibrational wave functions φ_{li}(R) and φ_{mf}(R) are the same. In Eq. (4.3) we can drop the index of the electronic state and use the usual orthonormality conditions for the vibrational functions of two different electronic states l and m

$$\int \varphi_{li}(R)\,\varphi_{mf}(R)\,dR = \delta_{if}.$$

Thus we arrive at a strict selection rule. It can be formulated as the rule: "For identical adiabatic potentials in the Condon approximation, all transitions involving a change of the vibrational state of the lattice are forbidden." For comparison with the selection rule for purely vibrational transitions (4.5), (4.6) can be written as

$$\Delta i = f - i = 0. \tag{4.7}$$

The selection rule (4.6) or (4.7) is never exactly obeyed, but it can be approximately true for systems with very small Stokes losses (for example, f-f transitions in rare-earth ions in crystals). Moreover, it is exact in the limit (in N^{-1}, where N is the number of band vibrations) in the very realistic case where an electronic transition in an impurity molecule interacts with band vibrations. We will see below that this leads to the appearance of a quasiline vibronic spectrum. The origin of the Mössbauer effect can also be understood on the basis of this special case where the selection rule (4.6) or (4.7) is valid in the limit. Later we will return to this effect, which is similar to quasiline vibronic spectra. We note that the Franck–Condon principle is applicable (with a somewhat different interpretation — in momentum space) to the Mössbauer effect also (see Appendix I).

We further note that the selection rules (4.6) or (4.7) are evidently closely related to the Condon approximation and do not depend directly (the dependence is "implicit" through the accuracy of the Condon approximation) on the kind of interaction with light, i.e., on whether dipole, quadrupole, or other transitions are under consideration. It is not difficult to modify the selection rule (4.6) to take account of deviations from the Condon approximation. For example, consideration of the linear terms in the expansion of the electronic matrix element (4.2) leads to a curious result — a rule $\Delta i = \pm 1$, i.e., to the usual selection rule (4.5) for a harmonic oscillator.

Exercise 2.   Calculate the selection rule for vibronic transitions with $W_l(R) = W_m(R)$, taking account of the linear and quadratic terms in the expansion (4.2) of the electronic matrix element. Use harmonic oscillator wave functions for the wave functions φ_i.

Exercise 3.   Explain how the purely vibrational transitions are related to the electronic dipole moment, i.e., to the first term in (4.1), in the same approximation.

Question.   Is it necessary that the vibrations be harmonic for the selection rule (4.6) to be valid?

Chapter 2

Theory of Quasiline Vibronic Spectra of Impurity Centers

§ 5. The Impurity Center Model

We consider a single impurity center (foreign molecule, impurity, or any other point defect) in a crystal. Such an approach is valid for low concentrations of impurity centers, where they can be regarded as independent of each other. We use the adiabatic approximation and therefore should make it clear how the electronic states will be treated.

We will assume that the ground and excited electronic states of the impurity do not enter into the composition of the electronic band states of the crystal. In other words, we assume that the impurity electrons are localized at the impurity in both the ground and excited states. Furthermore, we will treat the interaction of the impurity center electrons both with band vibrations and with localized vibrations.

The basic version of the theory (or the basic model) will involve the following additional assumptions:

(1) The transition probability is calculated in the first order of time-dependent perturbation theory.

(2) The wave functions of the system are constructed using the adiabatic approximation.

(3) The Condon approximation is used.

(4) The harmonic approximation is applicable to the lattice vibrations.

(5) When the electronic state changes, only the equilibrium positions of the normal oscillators change, i.e., we neglect changes in the interatomic force constants due to the electronic transition.

We will return later to various extensions of the basic model.

We will calculate the line shape as a function $I(\omega)$ which is given by

$$I(\omega) = A \sum_i v_i \sum_f |P_{fi}|^2 \delta (E_f - E_i - \hbar\omega),$$

where A is a frequency-independent constant, and v_i is the occupation probability of the i-th initial state (see Appendix V). Neglecting the frequency dependence of A is yet another simplifying assumption.

Before turning to the solution of the problem formulated above, we clarify certain generally-known concepts and terms.

§ 6.   The Energy of a Purely Electronic Transition, Zero-Phonon Transitions, and Quasilines

Among vibronic transitions a special place is occupied by the purely electronic transitions — transitions in which the vibrational state of the lattice (molecule) does not change. In Fig. 6 we show three purely electronic transitions in absorption, 0–0, 1–1, and 2–2, with energies E_{00}, E_{11}, and E_{22}. Evidently with different curvatures of the potential curves $W_I(R)$ and $W_{II}(R)$, due to differences in the vibrational quanta ($\hbar\omega_I \neq \hbar\omega_{II}$), the energies of these three purely electronic transitions will not be equal: $E_{00} \neq E_{11} \neq E_{22}$. To be explicit we will call $\varepsilon_0 = E_{00} + \frac{1}{2}\hbar(\omega_I - \omega_{II})$ the purely electronic energy, having in mind the characteristics of the given pair of potential curves (surfaces)* and not the ener-

---

* In the realistic case of a crystal with many vibrational degrees of freedom,

$\varepsilon_0 = E_{00} + \frac{1}{2}h\left(\sum_s \omega_{Is} - \sum_{s'} \omega_{IIs'}\right)$, where s and s' are the indices of the normal

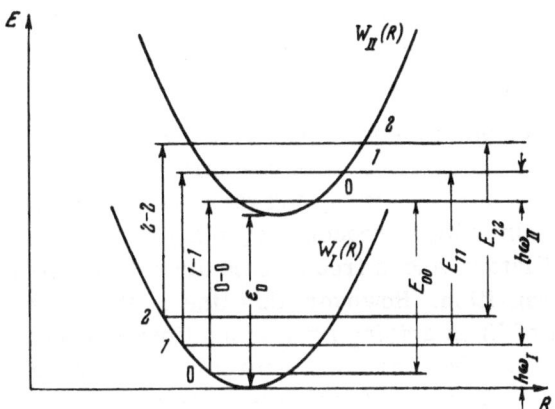

Fig. 6. Transitions $0-0$, $1-1$, and $2-2$, with no change of
the vibrational state (quantum number) of the oscillator.
If $h\omega_I \neq h\omega_{II}$, then the transition energies are different
($E_{00} \neq E_{11} \neq E_{22}$) and even the $0-0$ transition is accompanied by a change in the vibrational energy (of the zero-
point energy).

gies of all possible purely electronic transitions. Such a definition of the purely electronic energy evidently agrees with that generally used in the theory of luminescence. Note that the definition
of the purely electronic transition as a transition in which the vibrational energy of the lattice does not change is not quite exact
even when applied to a $0-0$ transition; the difference between the
zero-point energies of the oscillators, $\frac{1}{2} h\Sigma_s (\omega_{IIs} - \omega_{Is})$, corresponds
to vibrational motion of the lattice (s is the number of the normal
mode). This energy difference is not large but we will see that in
quasiline spectra there can be $0-0$ quasilines so narrow that the
influence of such fine effects is fully perceptible. In particular,
the isotope shift of a purely electronic line can be related to the
difference in the zero-point energies of the vibrations. A shift in
the Mössbauer line is also observed, which is related to the change
of the zero-point vibrational energy (the so-called second-order
Doppler shift).

Transitions without a change in the vibrational state of the
lattice are called z e r o - p h o n o n transitions since the lattice
modes in the ground and excited electronic states, and E_{00} is the energy of a transition
between the states in which all of the oscillators are in their ground states.

vibrations can be regarded as quasiparticles – phonons.* We could also arbitrarily call localized vibrations phonons (or "localized phonons").

If the vibration frequencies are the same in both electronic states then the energies of all of the zero-phonon transitions are equal ($E_{00} = E_{11} = E_{nn}$) (Fig. 6) and there is a single line in the vibronic spectrum with a frequency $\omega_0 = E_{00}/h$. This would occur for a set of oscillators whose frequencies did not change during the electronic transition. However, this line is always the superposition of a set of lines arising from transitions between different pairs of terms. Therefore it is appropriate to introduce the term quasiline.† The need for this term is especially clear if we bear in mind that, even for a small change in the oscillator frequency during the electronic transition, $E_{00} \neq E_{11} \neq E_{22}$, and the quasiline decomposes into its components. To find the internal structure of a quasiline spectrum, we ultimately need a special investigation. However, it is clear that in interpreting the shape of a quasiline and its temperature dependence it is quite important that it is not a line (a transition between a single pair of terms) but a quasiline. The quasiline reduces to a line at low temperatures ($h\omega \gg kT$). There are acoustic vibrations, however, whose frequencies begin at zero so that even liquid helium temperatures cannot be regarded as low. Therefore, strictly speaking, we always have quasilines in the vibronic spectra of crystals.

In those cases where it is not necessary to emphasize the quasiline nature of a line, we will use the simple term "line." In particular, instead of zero-phonon quasiline, we will often speak of the usual "purely electronic line," understanding it to be a quasiline which arises from the superposition of all possible zero-phonon transitions, i.e., the superposition of $0-0$, $1-1$, $2-2$, ... transitions (Fig. 6).

In a number of cases the zero-phonon quasiline in a vibronic spectrum is quite narrow. At low temperatures the half-widths

---

* In speaking of the excitation of the s-th band vibration from the i-th level into the f-th, we could use instead the language of quasiparticles and say that $f-i$ phonons are created (or annihilated) in the s-th band vibration.

† This term was first applied by Shpol'skii to the narrow lines which he observed in the vibronic spectra of impurity molecules in molecular crystals (the Shpol'skii effect, see [22]).

can be several tenths of a cm^{-1} and are basically determined by the lattice inhomogeneity.

It is important to understand that the zero-phonon line can be very narrow at low temperatures and in good crystals. The Doppler effect, in contrast to its effect on the spectral lines of atoms in a gas, does not broaden a quasiline. Therefore the widths of the narrowest lines of atomic spectra are not the limit for the widths of impurity quasilines in a crystal, for which the limiting width is the radiation width (in principle).

It is interesting to note that the narrowest* spectral line observed to date — the zero-phonon gamma line in the Mössbauer effect — is in fact a quasiline.

Exercise 4. Construct the scheme for the appearance of lines and quasilines in the dipole transitions of a harmonic oscillator. What effect does a small anharmonicity have on the lines, quasilines, and transition probabilities?

Exercise 5. Find the intensity ratios of the first three components of the quasiline in dipole absorption by a harmonic oscillator with a wave number $K = 200$ cm^{-1} for $T = 4$, 100, and 300 K.

§ 7. The Integrated Intensity of the Purely
Electronic Line and Its Temperature
Dependence in the Simplest Case

We will now calculate, for the basic model introduced in a previous section, the integrated intensity of the zero-phonon quasiline ("area under the quasiline") and its temperature dependence. We will consider an impurity center with no localized vibrations.

We consider a macroscopic crystal with a number N of normal modes which is enormous, of the order of Avogadro's number. Since we neglect the "scrambling" of the normal coordinates during the electronic transition (i.e., the change in the system of normal coordinates which arises when the interatomic force constants change) there is still a unique relation between the normal modes

---

* In the sense of the ratio $\Delta\omega/\omega$, where $\Delta\omega$ is the line width.

before and after the electronic transition* and we will describe
the vibrations in both electronic states by the same set (i.e., by
the number s) of oscillators. We will let the s-th normal mode be
in state i before the electronic transition and in state f after the
transition. For the lattice as a whole we have a transition from a
vibrational state $\{Ii_s\}$ for the initial electronic state I to a vibrational state $\{II f_s\}$ for the final electronic state II. The meaning
of this notation is as follows:

Oscillator (normal mode)
number $s = 1, 2, 3, \ldots, s, \ldots, N$

State number (energy
level of the oscillator) $\{i_s\} = i_1, i_2, i_3, \ldots, i_s, \ldots, i_N$

Example: One possible
set of oscillator states is $\{i_s\} = 2, 0, 3, \ldots, 7, \ldots, 1.$

$$(7.1)$$

The zero-phonon quasilines of interest to us are formed
from transitions in which the vibrational state of the lattice does
not change. Accordingly, we consider only those transitions in which
the vibrational quantum number of each oscillator is the same before and after the transition. Therefore in our problem it is logical
to choose for the final vibrational state of the lattice:

Oscillator number $s = 1, 2, 3, \ldots, s, \ldots, N$

State number $\{f_s\} = f_1, f_2, f_3, \ldots, f_s, \ldots, f_N$
such that $f_1 = i_1;\ f_2 = i_2;\ f_3 = i_3;\ f_s = i_s, \ldots, f_N = i_N$

Example: the set
of quantum numbers $\{f_s\} = \{i_s\} = 2, 0, 3, \ldots, 7, \ldots, 1.$
chosen in (7.1)
should be reproduced
$$(7.2)$$

According to the quantum-mechanical Franck–Condon principle the probability of a transition of the s-th oscillator from
state i_s into state $f_s = i_s$ is given in the Condon approximation by
the absolute square of the overlap integral of the wave functions
[See Eq. (4.3)].

---

* A correspondence continues to exist for band vibrations even with scrambling of the
  normal coordinates, but this requires a proof.

$$w\,(i_s \to i_s) = A_s \left| \int \varphi_{\text{II}\,i_s}(q_s)\,\varphi_{\text{I}\,i_s}(q_s)\,dq_s \right|^2, \tag{7.3}$$

where A_s is a constant, independent of the frequency.

Because of the independence of the normal modes, the zero-phonon transition probability for the set of oscillators (from a given initial state) can be expressed as the product of the zero-phonon transition probabilities (7.1) for all of the oscillators

$$W\,(\text{I}\,i_s \to \text{II}\,i_s) = \prod_s w\,(i_s \to i_s) = \prod_s A_s \left| \int \varphi_{\text{II}\,i_s}(q_s)\,\varphi_{\text{I}\,i_s}(q_s)\,dq_s \right|^2. \tag{7.4}$$

For example, for the initial state chosen in (7.1), (7.4) takes the explicit form

$$W\,(\text{I}\,i_s \to \text{II}\,i_s) = A_1 \left| \int \varphi_{\text{II}2_1}\varphi_{\text{I}2_1}\,dq_1 \right|^2 A_2 \left| \int \varphi_{\text{II}0_2}\varphi_{\text{I}0_2}\,dq_2 \right|^2 \cdots A_N \left| \int \varphi_{\text{II}1_N}\varphi_{\text{I}1_N}\,dq_N \right|^2,$$

where $\varphi_{\text{II}\,2_1}$ is the wave function of the first harmonic oscillator in state number 2 in electronic state II, i.e., the oscillator is excited into a vibrational level with quantum number 2.

To find the total integrated intensity of the zero-phonon quasiline, I(T), it is necessary according to (V.8) to sum the contributions from all possible initial states $\text{I}i_s$ of the lattice, taking into account the probability of each of these states. If we make the assumption, which is quite good for ordinary absorption and luminescence, that the vibrations are in thermal equilibrium before the transition, then the problem reduces to averaging over a Boltzmann distribution

$$I\,(T) = A\,\langle W\,(\text{I}\,i_s \to \text{II}\,i_s) \rangle_T =$$

$$A\,|D|^2 F\,(T) \sum_{[i_s]} \exp\left[-\frac{h}{kT} \sum_{\{i_s\}} \left(i_s + \frac{1}{2} \right) \omega_s^{\text{I}} \right]$$

$$\times \prod_{\{i_s\}} |\varphi_{\text{II}\,i_s}\varphi_{\text{I}\,i_s}\,dq_s|^2. \tag{7.5}$$

Here we have used the following notation: $D \equiv D_{\text{II}\,\text{I}}^0$ is the electronic matrix element (4.2b) in the Condon approximation, F(T) is the normalized Boltzmann distribution function, and ω_s^{I} is the frequency of the s-th normal mode in electronic state I. Because of the narrowness of the quasiline it is usually valid to neglect the frequency dependence of the factor A. Use of $\{i_s\}$ with a sum (product) sign denotes a sum (product) over a fixed set of vibrational states i_s, and $[i_s]$ denotes a sum over all poss-ible sets i_s (the index I of the electronic state will be dropped here and in what follows when it does not cause confusion). The sign $< \ldots >_T$ denotes a thermal equilibrium average.

We will now calculate $I(T)$, beginning with the overlap inte-gral. We deal only with band vibrations. In Section 3 we showed that the equilibrium positions of the oscillators change by an amount proportional to $N^{-1/2}$, where N is the number of normal modes. The changes in the oscillator frequencies and axes can be neglected for now. Then the vibrational wave functions for the same s-th normal mode in the two different electronic states of the impurity are related as follows

$$\varphi_{II i_s}(q_s) = \varphi_{I i_s}(q_s - q_{s0}), \qquad (7.6)$$

where the shift in the equilibrium position is $q_{s0} = a_s N^{-1/2}$. The coefficients depend on the nature of the multidimensional adiabatic potential and its change in a given electronic transition. In general it is as difficult to calculate them as it is to solve the electronic equation. They will enter the formulas as parameters of the the-ory which are to be found from experiment or to be determined by a special calculation.

The shift in the equilibrium position will in fact be regarded as infinitesimal, since we let N go to infinity. But when N in-creases, the number of normal modes with which the electronic transition of the impurity center can interact also increases, and therefore it is not correct to set q_{s0} equal to zero.

We expand the wave function (7.6) in a series in q_{s0} and re-tain only the first three terms (the remaining terms decrease more rapidly than N^{-1}, and it is not difficult to see that they do not give a finite contribution to the transition probability)

$$\varphi_{II i_s}(q_s) = \varphi_{I i_s}(q_s - q_{s0}) = \varphi_{I i_s}(q_s) - \frac{d\varphi_{I i_s}}{dq_s} q_{s0} + \frac{1}{2} \frac{d^2\varphi_{I i_s}}{dq_s^2} q_{s0}^2 + \ldots \quad (7.7)$$

We thus have for the overlap integrals in (7.4)

$$J_{i_s} = \int_{-\infty}^{+\infty} \varphi_{II i_s} \varphi_{I i_s} dq_s = \int_{-\infty}^{+\infty} \varphi_{I i_s} \varphi_{I i_s} dq_s \; -$$

$$-\frac{a_s}{\sqrt{N}}\int_{-\infty}^{+\infty}\varphi_{1i_s}\frac{d}{dq_s}\varphi_{1i_s}dq_s+\frac{1}{2}\frac{a_s^2}{N}\int_{-\infty}^{+\infty}\varphi_{1i_s}\frac{d^2}{dq_s^2}\varphi_{1i_s}dq_s+\ldots \qquad (7.8)$$

The first term on the right-hand side is unity because the vibrational functions are normalized. The second term is zero since differentiating a harmonic oscillator wave function changes its parity and one gets an integral of an odd function between symmetric limits. The third term is easily calculated using the fact that the kinetic energy operator $\hat{T}$ of the s-th oscillator takes the form

$$\hat{T}=\frac{1}{2m_s}\hat{p}^2=-\frac{h^2}{2\,m_s}\frac{d^2}{dq_s^2}\,. \qquad (7.9)$$

It is necessary to comment on the meaning of the mass m_s. Since we speak of normal modes, the numerical value of m_s, strictly speaking, is not generally defined. One speaks of the frequency of the normal mode. The force constant and the mass are not separately defined, as can easily be seen from Appendix II.

In particular, we see from (II.12) that the cartesian displacement of the k-th atom in a normal mode contains the atomic mass as the factor $m_k^{1/2}$. We have introduced m_s on the basis of dimensional considerations to preserve the usual dimensionality of all the quantities entering the formulas. In order to obtain the form of an equation in normal coordinates as usually given in the literature, one must set all of the $m_s = 1$ in our equation, replace the cartesian displacements by the weighted displacements $u_k = m_k^{1/2}x_k$, and change the interpretation and dimensionality of the other quantities correspondingly. This form of the displacements is used in section 5 of Appendix II.

With (7.9) in mind, we can express the integral of interest in terms of the average kinetic energy (dropping the electronic state index I)

$$\frac{1}{2}\int\varphi_{i_s}\frac{d^2}{dq_s^2}\varphi_{i_s}dq_s=-\frac{m_s}{h^2}\int\varphi_{i_s}\hat{T}\varphi_{i_s}dq_s=-\frac{m_s}{h^2}\langle\hat{T}\rangle_{i_s}\,. \qquad (7.10)$$

where $\langle\ldots\rangle_{i_s}$ denotes quantum-mechanical averaging over the i-th state of the s-th oscillator. From the virial theorem (which also holds in quantum mechanics) it is well known that the average kinetic energy of a harmonic oscillator is half of its total energy

$$\langle \hat{T} \rangle_{i_s} = \frac{1}{2} E_{i_s} = \frac{1}{2} \left(i_s + \frac{1}{2} \right) \hbar \omega_s. \tag{7.11}$$

Thus the overlap integral is

$$J_{i_s} = 1 - \frac{a_s^2 m_s}{2 \hbar^2 N} E_{i_s}. \tag{7.12}$$

We have for its square (neglecting terms of order $1/N^2$)

$$J_{i_s}^2 = 1 - \frac{a_s^2 m_s}{\hbar^2 N} E_{i_s}. \tag{7.13}$$

We now turn to the average over initial states in Eq. (7.5). It follows from the independence of the normal modes that the average can be carried out for each oscillator separately. Actually, (7.5) can be written as

$$I(T) = A |D|^2 F_1(T) \sum_{i_1=0}^{\infty} \exp\left[-(\hbar\omega_1/kT) \left(i_1 + \frac{1}{2} \right) \right]$$

$$\times J_{i_1}^2 F^*(T) \sum_{\{i_s^*\}} \exp\left[-(\hbar/kT) \sum_{\{i_s^*\}} \left(i_s + \frac{1}{2} \right) \omega_s \right] \times \prod_{\{i_s^*\}} J_{i_s}^2, \tag{7.14}$$

where the asterisks on the indices i_s and the normalizing factor $F(T)$ mean that we have excluded oscillator number 1 from the set. With this notation the averaging over the states of oscillator number 1 can easily be carried out.

$$F_1(T) \sum_{i_1=0}^{\infty} \exp\left[-(\hbar\omega_1/kT)\left(i_1 + \frac{1}{2} \right) \right]\left[1 - \frac{a_1^2 m_1}{N\hbar^2} \hbar\omega\left(i_1 + \frac{1}{2} \right) \right] = 1 - \frac{a_1^2 m_1}{N\hbar^2} E_1(T). \tag{7.15}$$

Here $E_1(T)$ is the average thermal energy of a linear harmonic oscillator

$$E_1(T) = \hbar\omega_1 \left(\bar{i}_1 + \frac{1}{2} \right) = \frac{1}{2} \hbar\omega_1 \coth \frac{\hbar\omega_1}{2kT}, \tag{7.15a}$$

where $\bar{i}_1$ is the average quantum number of the oscillator or the average number of phonons in band vibration number 1.

We can now separate out the states of oscillator number 2 from the remaining set $\{i_s^*\}$ of (7.14) and average over them. We then separate out the third oscillator, etc. We find finally

$$I(T) = A |D|^2 \prod_{s=1}^{N} \left(1 - \frac{a_s^2 m_s \omega_s}{2Nh} \coth \frac{\hbar \omega_s}{2 kT} \right). \tag{7.16}$$

We now transform this product, expanding its logarithm in series

$$\ln \prod_{s=1}^{N} \left(1 - \frac{1}{N} \frac{a_s^2 m_s}{h^2} E_s(T) \right) = \sum_{s=1}^{N} \ln \left(1 - \frac{1}{N} \frac{a_s^2 m_s}{h^2} E_s(T) \right) = - \frac{1}{N} \sum_{s=1}^{N} \frac{a_s^2 m_s}{h^2} E_s(T) \tag{7.17}$$

(keeping in mind that N will go to infinity, we can regard the latter equality as exact).

We finally find for the integrated intensity of the zero-phonon line (with no localized vibrations)

$$I(T) = A|D|^2 \exp \left[-\sum_{s=1}^{N} \frac{m_s q_{0s}^2 \omega_s}{h} \left(\bar{i_s} + \frac{1}{2} \right) \right] \equiv A|D|^2 \exp \left[-\sum_{s=1}^{N} \frac{m_s q_{0s}^2}{h^2} k \tau_s \right]. \tag{7.17a}$$

The average number of phonons $\bar{i_s}$ goes to zero when $T \to 0$ and in the high temperature limit ($kT \gg \hbar \omega_s$), where classical statistics are valid, $\bar{i_s} + \frac{1}{2} = kT/\hbar \omega_s$. Sometimes we will also use the "effective temperature" τ_s of a harmonic oscillator

$$k \tau_s \equiv E_s(T) \equiv \left(\bar{i_s} + \frac{1}{2} \right) \hbar \omega_s = \frac{\hbar \omega_s}{2} \coth \frac{\hbar \omega_s}{2 kT}, \tag{7.17b}$$

which is convenient in the analysis of temperature dependences. At high temperatures ($kT \gg \hbar \omega_s$), τ_s goes over to the usual temperature: $\tau_s \to T$. At low temperatures ($kT \ll \hbar \omega_s$), $k \tau_s$ goes over to the zero-point energy of the oscillator, $k \tau_s \to \hbar \omega_s / 2$.

We will now analyze the results obtained. We rewrite the exponent in a somewhat different form:

$$I(T) = A|D|^2 \exp \left[-\sum_{s=1}^{N} \frac{2 \mathscr{P}_s}{\hbar \omega_s} \left(\bar{i_s} + \frac{1}{2} \right) \right]$$

$$\equiv A|D|^2 \exp \left[-\sum_{s=1}^{N} \frac{2 \mathscr{P}_s}{\hbar \omega_s} \frac{k \tau_s}{\hbar \omega_s} \right]. \tag{7.18}$$

Here we have introduced the notation $\mathscr{P}_s \equiv \frac{1}{2} m_s \omega_s^2 q_{s0}^2$. The advantage of this form is that the exponent now contains only two dimensionless ratios of energies, and the vibrational quantum energy plays the role of a natural unit of measurement.

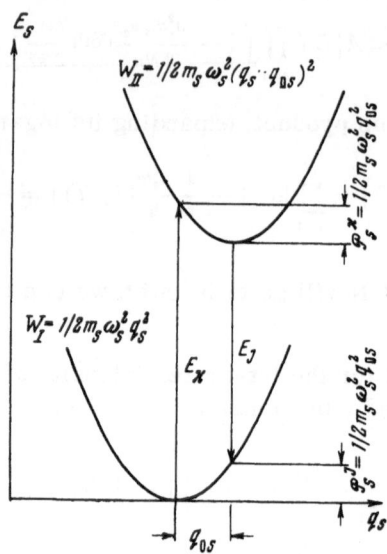

Fig. 7. Stokes energy loss $\mathscr{P}_s$ and its dependence on the shift in the equilibrium
position and the interatomic force constant
$a = \frac{1}{2}m\omega^2$; for harmonic oscillators with
equal force constants $\mathscr{P}_s^\varkappa = \mathscr{P}^J$.

To explain the physical meaning of $\mathscr{P}_s$ we go back to the potential curve diagram of the s-th oscillator. In this case the potential curves are a pair of parabolas shifted with respect to each
other along the coordinate (q_s) axis. The relationships between the
parameters are shown in Fig. 7. We see that $\mathscr{P}_s$ is characteristic
of the mutual positions of the potential curves (in the present case
the shapes of the curves are the same). The energy $2\mathscr{P}_s$ is the
difference in length between the "energy arrows" $E_\varkappa$ and E_J. The
quantity $E_\varkappa(E_J)$ is close to the energy at which the maximum of the
vibronic absorption (emission) band lies for diatomic molecules
having the potential curves shown, in vibrational thermal equilibrium. The difference $E_\varkappa - E_J$ is approximately the Stokes shift between the maxima of the absorption and emission bands. We will
call $\mathscr{P}_s = \frac{1}{2}(E_\varkappa - E_J)$ the Stokes energy loss.

This definition (with the factor $\frac{1}{2}$) is convenient because
with different frequencies $\omega_I \neq \omega_{II}$ of the potential curves we have
$E_\varkappa - E_J = \mathscr{P}^\varkappa + \mathscr{P}^J$, for the Stokes shift, where

$$\mathscr{P}^x \equiv \frac{1}{2} m_s \omega_{11}^2 q_{s0}^2, \; \mathscr{P}^J \equiv \frac{1}{2} m_s \omega_1^2 q_{s0}^2 . \qquad (7.19a)$$

The Stokes loss in emission (absorption) $\mathscr{P}^J$ $(\mathscr{P}^x)$ is the average energy imparted to vibrations in a phototransition.

We further note that the distribution of the purely electronic transition energy (i.e., the energy difference between the minima of the multidimensional potential surfaces) among normal oscillators is completely arbitrary. Therefore, the quantities E_x and E_J do not make sense for a single normal mode. Only their difference $2\mathscr{P}_s$ has significance. The potential curve diagram of Fig. 7 and the discussion of the vibronic bands and vibrational relaxation for a single normal mode are arbitrary and are only used to explain the use of the term "Stokes loss" for $\mathscr{P}_s$.

We now consider (7.18) for two temperature limits.

1. Low temperatures ($kT \ll \hbar\omega_s^{min}$, where ω_s^{min} is the lowest vibration frequency which interacts with the electronic transition).[*]

In this case

$$k\tau_s \rightarrow \frac{\hbar\omega_s}{2} ,$$

$$I(T \rightarrow 0) = A|D|^2 \exp \left[-\sum_{s=1}^{N} \frac{\mathscr{P}_s}{\hbar\omega_s} \right] = A|D|^2 \exp \left[-\sum_{s=1}^{N} p_s \right] . \qquad (7.20)$$

The quantities

$$p_s \equiv \frac{\mathscr{P}_s}{\hbar\omega_s} \qquad (7.19b)$$

will be called the Stokes losses, expressed in terms of the number of vibrational quanta, or the dimensionless Stokes losses.

Thus in the low-temperature limit the integrated intensity of the purely electronic line, I(T), has a finite value which is greater the lower the total Stokes loss (summed over all normal

---

[*] Strictly speaking, when $N \rightarrow \infty$ the lowest acoustic vibration frequency goes to zero and the introduction of a finite ω_s^{min} is not justified. It is natural to assume, however, that some portion of the lowest-frequency acoustic phonons do not generally interact with the electronic transition, i.e., $\mathscr{P}_s = 0$ for them. These acoustic frequencies correspond to vibrations with long wavelengths which actually only "feel" the change of the electronic state of the impurity to an extremely small degree.

modes). The factor $|D|^2$ is common to all of the vibronic bands. Equation (7.20) can also be written as

$$I(T \to 0) = A|D|^2 \exp\left[-\frac{1}{\hbar\bar{\omega}}\sum_{s=1}^{N}\mathcal{P}_s\right] = A|D|^2 \exp\left[-\frac{1}{\hbar\bar{\omega}}\mathcal{P}\right]. \quad (7.21)$$

Here $\bar{\omega}$ is a certain average frequency of the band vibrations and $\mathcal{P} \equiv \sum_s \mathcal{P}_s$ is the total Stokes energy loss, which is the analog of $\mathcal{P}_s$ for multidimensional potential surfaces. The latter expression for $I(T \to 0)$ does not contain any explicit dependence on the normal modes.

We know little about $\bar{\omega}$. To determine it one must calculate the sum in (7.20). However, it is evident that $\bar{\omega}$ is greater the larger the vibrational quanta which interact effectively with the electronic transition. With a given Stokes loss energy, $I(T \to 0)$ is greater the larger $\bar{\omega}$ is. In general, there is an effective interaction with oscillators of various frequencies and then it is most correct to use the dimensionless Stokes losses p_s. Sometimes it is evidently appropriate to introduce several frequencies $\bar{\omega}_r$ (r is the vibrational branch index). We can use the Debye frequency as a crude estimate of $\bar{\omega}$, or the limiting frequencies of the optical branches, which are better for the theory of vibronic transitions.

2. High Temperatures ($kT \gg \hbar\omega_s^{\max}$). In this case the effective temperature τ_s can be replaced by the temperature T and we find from (7.18) for the integrated intensity of the purely electronic line

$$I(kT \gg \hbar\omega_s^{\max}) = A|D|^2 \exp\left[-kT\sum_{s=1}^{N}\frac{2\mathcal{P}_s}{(\hbar\omega_s)^2}\right] = A|D|^2 \exp\left[-\frac{kT}{\hbar\bar{\bar{\omega}}}\frac{2\mathcal{P}}{\hbar\bar{\bar{\omega}}}\right] \quad (7.22)$$

Evidently the "certain average" $\bar{\bar{\omega}}$ of the frequencies introduced here will as a rule differ from the $\bar{\omega}$ introduced previously. It can also be estimated from a limiting optical frequency or the Debye frequency.

We see that at high temperatures the intensity of the zero-phonon line decreases exponentially with increasing temperature, and that it drops more rapidly the larger the Stokes loss and the lower the average vibrational quantum, $\hbar\bar{\bar{\omega}}$.

In the intermediate temperature range the temperature dependence is a complicated function, but it is not difficult to see that the basic tendency is a drop of I(T) with temperature in which I(T) is less the larger the total Stokes loss in units of the vibrational quanta.

Consequently the purely electronic line is intense in systems with low Stokes losses and at low temperatures. In systems with large Stokes losses, for example in alkali halide crystals of the KCl-Tl type, in which the ratio $\mathscr{P}/\hbar\bar\omega$ is several tens, it is extremely weak. I(T) attains a significant value in molecular crystals containing impurities (Shpol'skii systems) and in crystals activated with rare earth ions ("laser crystals"). Evidently in these systems the ratio $\mathscr{P}/\hbar\bar\omega \approx 0.1 - 1.0$.

A knowledge of the integrated intensity of the zero-phonon quasiline by itself is not sufficient to form an opinion about that part of the vibronic band which contains this quasiline. It is also necessary to know the position of the purely electronic line. Furthermore, with a given integrated intensity the line shape can differ. Therefore it is also necessary to know the half-width of the purely electronic quasiline. Moreover, with respect to other transitions (which are accompanied by creation and annihilation of phonons and which contribute in the actual frequency range around the purely electronic transition energy) it is necessary to know, if only in general terms, the nature of the spectrum which they produce. These transitions will be discussed in section 10. We now turn to the characteristic position and width of the purely electronic quasiline itself.

Since in the simplest model of the vibrational sublevels which we are considering here, they have no widths, and the changes in the oscillator frequencies are neglected, it is quite clear (for example, from consideration of the potential curve diagram of Fig. 6) that the purely electronic line, independent of temperature, lies at the frequency of the purely electronic transition, $\omega_0 = E_{00}h^{-1}$, and is of zero width. We can also show this mathematically.

Exercise 6.   Calculate the normalizing factor $F_1(T)$ in Eq. (7.14).

§ 8.   Position, Width, and Peak Intensity

of the Purely Electronic Line in the

Simplest Case

We will characterize the position and width of the purely electronic line by the first moments of the intensity distribution function for this quasiline (see Appendix VI for a discussion of the method of moments). As is well-known, the zeroth moment gives the area under the distribution curve, the normalized first moment (i.e., the first moment divided by the zeroth) is the coordinate of the "center of gravity" of the distribution, and the normalized second central moment (i.e., the moment calculated relative to the center of mass and divided by the zeroth moment) characterizes the width of the distribution.

We can find the intensity distribution in the purely electronic line from the general definition of the l-th order moment S_l of the transition probability distribution W with respect to transition energy (see, for example, [8, 68])

$$S_l = \sum_i v_i \sum_f (E_{\mathrm{II}f} - E_{\mathrm{I}i})^l W_{if}, \tag{8.1}$$

(v_i is the occupation probability of the i-th initial state, W_{if} is the probability of the transition $i \to f$, and E_f and E_i are the energies of the i-th initial and f-th final states) if in the sum over final states f we retain only those terms which correspond to zero-phonon transitions, i.e., $f = i$. We also assume that in describing vibrations in the harmonic approximation we have $E_{\mathrm{I}i} = \sum_s \hbar \omega_{\mathrm{I}s} \left(i_s + \frac{1}{2} \right)$ and $E_{\mathrm{II}f} = \varepsilon_0 + \sum_s \hbar \omega_{\mathrm{II}s} \left(f_s + \frac{1}{2} \right)$, where ε_0 is the energy of the purely electronic transition, s is the normal mode index, and the indices I and II refer to the initial and final electronic states. We find a formula for the l-th moment S_l^0 of the purely electronic line (which is also valid with unequal frequencies $\omega_{\mathrm{I}s} \neq \omega_{\mathrm{II}s}$):

$$S_l^0 = S_{l_{\text{with } i=f}} = \sum_i v_i \sum_f \left[\varepsilon_0 + \sum_s \hbar \omega_{\mathrm{II}s} \left(f_s + \frac{1}{2} \right) - \right.$$

$$-\sum_s \hbar\omega_{\mathrm{Is}}\left(i_s + \frac{1}{2}\right)\Big]^l W\,(\mathrm{I}\,i_s \to \mathrm{II}\,f_s)\,|_{i=f}$$

$$=\sum_i v_i\left\{\varepsilon_0 + \sum_s h\left[\omega_{\mathrm{IIs}}\left(i_s + \frac{1}{2}\right) - \omega_{\mathrm{Is}}\left(i_s + \frac{1}{2}\right)\right]\right\}^l W\,(\mathrm{I}i_s \to \mathrm{II}i_s). \qquad (8.2)$$

In the simplest model which we have considered, $\omega_{\mathrm{IIs}} = \omega_{\mathrm{Is}}$ and Eq. (8.2) can be greatly simplified

$$S_l^0 = \varepsilon_0^l \sum_i v_i W\,(\mathrm{I}i_s \to \mathrm{II}i_s) = \varepsilon_0^l S_0^0. \qquad (8.3)$$

The zeroth moment of the purely electronic line in the simplest model is none other than the integrated intensity $S_0^0 = I(T)$ calculated in the previous section. Finally, we find for the l-th moment of the intensity distribution

$$S_l^0 = \varepsilon_0^l S_0^0 = \varepsilon_0^l I\,(T), \qquad (8.4)$$

where I(T) is defined by Eq. (7.17) [or (7.18)]. Equation (8.4) contains all of the information about the intensity, position, and shape of the purely electronic line. Its integrated intensity, which as previously noted is given by the zeroth moment, is I(T). We have for the first normalized moment

$$s_1 = \frac{S_1^0}{S_0^0} = \varepsilon_0, \qquad (8.5)$$

i.e., the "center of gravity" of the purely electronic line lies exactly at the energy of the purely electronic transition. The position of the "center of gravity" is independent of the temperature.

The second moment is

$$S_2^0 = \varepsilon_0^2 S_0^0, \qquad (8.6)$$

whence we have for the second central moment

$$\bar{s}_2 = \frac{1}{S_0^0}\,(S_2^0 - S_0^0 s_1^2) = 0. \qquad (8.7)$$

This result shows that the intensity distribution has zero width.*

---

* It is not difficult to see that the higher central moments S_l ($l = 3, 4, \ldots$) are also zero, as they should be if $\bar{s}_2 = 0$.

We ultimately conclude that in our model the purely electronic line is a line of zero width located at the frequency $\omega_0 = \varepsilon_0 h^{-1}$ and having a finite intensity I(T) which depends on the temperature. Using the δ-function we can write its shape function $I(\omega, T)$ as

$$I(\omega, T) = \delta(\omega - \omega_0) I(T), \qquad (8.8)$$

where I(T) is given by Eq. (7.17).

It is clear that in reality the line will have a finite width. But, since the approximations made in the simplest model are fairly good, we would expect that the purely electronic quasiline is intrinsically very narrow and has a very high peak intensity (i.e., the intensity at the maximum) for reasonably large I(T). If we reconsider all of the simplifications made in obtaining (8.8) it is clear that there is only one important point in which we must make corrections if we want to gain a correct idea of the line shape (the remaining assumptions may be too crude in some respects in a particular case, but in principle they can be realized exactly).

The width of a vibrational level is an important question: because the excited electronic state has a finite lifetime for spontaneous emission of light, the width of a vibrational sublevel in an excited electronic state cannot be less than $\Delta\varepsilon_0 = h\tau_{opt}^{-1}$, where τ_{opt} is the optical lifetime. In the presence of radiationless transitions, which decrease the lifetime $\tau < \tau_{opt}$ ($\tau_{opt}^{-1} + \tau_T^{-1} = \tau^{-1}$, where τ_T is the lifetime for radiationless transitions) $\Delta\varepsilon_0$ increases. The quantity $\Delta\varepsilon_0$ gives the finite frequency width of the excited electronic level, $\Delta\omega_{el} = \Delta\varepsilon_0 h^{-1} = 2\pi\tau^{-1}$. This quantity is in principle the lower limit $\Delta\omega_{min} = \Delta\omega_{el}$ for the widths of all vibrational levels of the excited electronic state and in principle is the lower limit for the observed width of the purely electronic line. The quantity $\Delta\omega_{min}$ corresponds to an extremely narrow spectral line. For example, with the lifetime $\tau \approx 3 \cdot 10^{-7}$ sec usual for allowed transitions we have $\Delta\omega_{min} \approx 2 \cdot 10^7$ sec^{-1}, or, if we express the level width in terms of wave numbers, $\Delta k_{min} \approx 10^{-3}$ cm^{-1}; for forbidden transitions with $\tau \approx 3 \cdot 10^{-4}$ sec the estimate gives $\Delta k_{min} \approx 10^{-6}$ cm^{-1}.

Clearly a very narrow spectral line is extremely sensitive to small interactions which could give an observable shift and greater relative broadening. Therefore it is of greatest value in the theory of quasilines to consider factors which are small corrections in the theory of broad vibronic bands. Recently, refined versions of the theory have been worked out. It turns out that the "scrambling" of the normal coordinates during the electronic transition leads to a substantial thermal broadening of the purely electronic quasiline. At low temperatures the observed width of the quasiline (which exceeds $\Delta\omega_{min}$ by two or three orders of magnitude) arises from small inhomogeneities in the crystal structure. The results of this work will be presented in Chapter IV. Here we further stress that the basic assumptions of the simple model which we have treated are reasonable, and that narrow purely electronic lines are actually observed in the spectra of crystals containing impurities. Although their widths are substantially greater than Δk_{min}, they are all very narrow and have very high peak intensities. Therefore it makes sense to extend the theory for the simplest model. The conclusions obtained are the basis for the interpretation of experimental data, and the starting point for the refined versions of the theory.

§ 9. The Integrated Intensity of a Vibronic Band

We now know the intensity, position, and shape of the purely electronic line and also its temperature dependence. In order to explain the properties and temperature dependence of the rest of the spectrum connected with a given electronic transition we first solve an auxiliary problem — the integrated intensity of the whole vibronic band, $S_0(T)$. We must calculate the zeroth moment $S_0(T)$ of the intensity distribution of all of the transitions corresponding to the electronic transition I → II from Eq. (8.1). This calculation can be simply and elegantly carried out in general form.* $S_0(T)$ is the sum of the intensities of all possible vibronic transitions corresponding to the given electronic transition I → II

$$S_0(T) = \sum_i n_i \sum_f W(\mathrm{I}\,i \to \mathrm{II}\,f), \qquad (9.1)$$

_____

* This calculation, as with the calculations of other moments, is related to spectroscopic sum rules [69] (see Appendix VI for more on the method of moments).

where n_i is the number of systems in the i-th vibrational level of the initial electronic state and the summation over f (in contrast to (8.2)) runs over all values of f (including $f = i$). The summation over i with weights n_i signifies an averaging over vibrational sublevels in the initial electronic state and summation over all indices f corresponds to considering all possible transitions — both zero–phonon and phonon transitions. But we do not make any additional simplifying assumptions about the nature of the vibrations or the distribution of the systems over the initial vibrational levels. The obtained result (9.2) is valid not only for the simplest model but persists for all models based on the adiabatic approximation. The final answer (9.3) is valid if, in addition to the adiabatic approximation, we also introduce the Condon approximation. In particular, the vibrations may be strongly anharmonic. The distribution over initial levels can also differ strongly from the thermal equilibrium distribution.

If we do not introduce the Condon approximation we have (here R is an abbreviated notation for the set of all coordinates describing the vibrations)

$$
S_0(T) = \sum_i n_i \sum_f \left| \int \varphi^*_{\mathrm{II}f}(R) M(R) \varphi_{\mathrm{I}i}(R) \, dR \right|^2
$$

$$
= \sum_i n_i \sum_f \int \varphi^*_{\mathrm{II}f}(R) M(R) \varphi_{\mathrm{I}i}(R) \, dR \int \varphi_{\mathrm{II}f}(R') M^*(R') \varphi^*_{\mathrm{I}i}(R') \, dR'
$$

$$
- \sum_i n_i \iint dR dR' M(R) M(R') \varphi_{\mathrm{I}i}(R) \, \varphi^*_{\mathrm{I}i}(R') \sum_f \varphi^*_{\mathrm{II}f}(R) \varphi_{\mathrm{II}f}(R')
$$

$$
= \sum_i n_i \iint dR dR' M(R) M(R') \varphi^*_{\mathrm{I}i}(R) \, \varphi_{\mathrm{I}i}(R') \delta(R - R')
$$

$$
= \sum_i n_i \int dR \, |M(R)|^2 \, |\varphi_{\mathrm{I}i}(R)|^2, \tag{9.2}
$$

where $M(R)$ is the electronic matrix element which is given by Eq. (4.2a) in the dipole approximation.

Here we have used the properties of the complete set of functions $\varphi_{\mathrm{II}f}(R)$, so that we have the relation (see, for example [70])

$$
\sum_f \varphi^*_{\mathrm{II}f}(R) \varphi_{\mathrm{II}f}(R') = \delta(R - R'),
$$

and the property of the δ–function

$$\int f(R') \delta(R - R') dR' = f(R).$$

We now introduce the Condon approximation $M(R) = M =$ const. It follows from (9.2) that

$$S_0(T) = \sum_i n_i |M|^2 \int |\varphi_{1i}(R)|^2 dR = N|M|^2 = \text{const},\qquad(9.3)$$

where N is the total number of systems (impurity centers in the crystal). We have arrived at a remarkable result: The integrated intensity of the whole spectral band, $S_0(T)$, corresponding to a single electronic transition, is independent of the distribution function n_i in the Condon approximation, i.e., it does not depend on the temperature. There is a peculiar conservation of the area under the spectral curve.* Deviations from the Condon approximation cause a change in the area of the band with temperature. For electronic transitions allowed by spatial symmetry these changes, as shown by experiment, are small [71].

It follows from the results obtained that the drop in the intensity of the zero-phonon quasiline with increasing temperature should be accompanied by a growth of intensity in other parts of the spectrum. Because of the conservation of the area under the whole band, the area disappearing from under the purely electronic quasiline should appear in other parts of the vibronic band.

Exercise 7. Estimate the fraction of integrated intensity $I(T)/S_0(T)$ lying in the purely electronic line with Stokes losses of $\mathscr{P} = 0.01$ eV, 0.1 eV, and 1.0 eV, effective frequencies $\bar{\omega}/c = \bar{\bar{\omega}}/c = 200$ cm^{-1}, and temperatures of 0, 50, 100, and 300 K. Show the results graphically.

§ 10. One-Phonon Transitions

We now turn to transitions which are accompanied by the creation or annihilation of a certain number of band vibration phonons. Our aim will be to see what else besides the purely elec-

---

* Remember that in the expression for the shape of the spectrum there is still a weakly frequency-dependent factor (see Appendix V). While it can be regarded as constant for the narrow purely electronic line to a high degree of precision, for the band as a whole there is a small correction so that the area of the band depends somewhat on the temperature even if the Condon approximation is valid. For absorption bands this dependence is very weak, while for emission bands it is somewhat more significant.

tronic line appears in the spectrum. In particular, it is important
to know how the spectrum looks in the vicinity of the purely electronic line. It might happen, for example, that phonon transitions
give other intense peaks in the immediate neighborhood of the
purely electronic line, which could merge with it into one broad
band.

We first consider one–phonon transitions, i.e., electronic
transitions accompanied by the creation or annihilation of one band
vibration phonon. We show that if deviations from the selection
rule (4.6) are small, the one–phonon transition probability is
greater the greater the excitation in the initial vibrational level.
It will likewise be shown that with increasing temperature, where
higher vibrational levels become occupied, the one–phonon transition probability increases and we get a qualitative answer to the
question posed at the end of the previous section: What happens
to the intensity of transitions in which the vibrational state
changes as the temperature increases?

For this purpose we go back to an investigation of the corresponding Franck–Condon factor of a harmonic oscillator with
unchanged frequency (the oscillator index is dropped). For a
transition in which one phonon is created, i.e., in a transition
where $i \to i + 1$, we have

$$J_{i,i+1} = \int \varphi_{1i}(q)\, \varphi_{11i+1}(q)\, dq, \tag{10.1}$$

where

$$\varphi_{11i+1}(q) = \varphi_{1i+1}(q - q_0).$$

After expanding in powers of q_0 and substituting in the integral we find

$$J_{i,i+1} = \int \varphi_i(q) \left[\varphi_{i+1}(q) + \frac{d}{dq}\varphi_{i+1}(q)\, q_0 + \dots \right] dq$$

$$= -q_0 \int \varphi_i(q)\, \frac{d}{dq}\varphi_{i+1}(q)\, dq = -q_0 \sqrt{\frac{(i+1)m\omega}{2h}}. \tag{10.2}$$

Here we have taken the orthogonality and normalization conditions for vibrational functions belonging to the same electronic
state into account

$$\int \varphi_i(q)\, \varphi_k(q)\, dq = \delta_{ik}, \tag{10.3}$$

and used the following well-known relations between the eigenfunctions of a harmonic oscillator

$$\bar{q}\frac{d}{dq}\varphi_i(q) = \frac{q}{\bar{q}}\varphi_i(q) - \sqrt{2(i+1)}\,\varphi_{i+1}(q), \qquad (10.4a)$$

$$\frac{q}{\bar{q}}\varphi_i(q) = \sqrt{\frac{i+1}{2}}\,\varphi_{i+1}(q) + \sqrt{\frac{i}{2}}\,\varphi_{i-1}(q), \qquad (10.4b)$$

where $\bar{q} = \sqrt{h/m\omega}$.

We thus have for a transition $i \to i+1$ in one individual oscillator*

$$W(i \to i+1) = |J_{i,i+1}|^2 = \frac{i+1}{2h}q_0^2 m\omega. \qquad (10.5)$$

When one phonon is annihilated in one individual oscillator we have similarly for the transition probability

$$W(i \to i-1) = |J_{i,i-1}|^2 = \frac{i}{2h}q_0^2 m\omega \qquad (10.6)$$

Note that both of the one-phonon transition probabilities in a single oscillator increase with increasing initial vibrational excitation. Since ever higher initial vibrational levels i are occupied with increasing temperature, the one-phonon transition probability increases with temperature.

We further recall that for the s-th band oscillator the shift in the equilibrium position is $q_{s0} = a_s N^{-1/2}$, where N is the number of vibrational degrees of freedom of the crystal and a_s is a finite constant. Since $W(i \to i+1) \sim q_{s0}^2 \sim N^{-1}$, both of the one-phonon transition probabilities are infinitesimals of order N^{-1} as $N \to \infty$. The corresponding spectral lines have intensities of the same order of magnitude. In one-phonon transitions corresponding to creation of a single phonon in the s-th mode, a line appears with a frequency $\omega_0 \pm \omega_s$, where ω_0 and ω_s are the frequencies of the purely electronic line and the s-th normal mode; the plus sign is to be taken for absorption and the minus sign for emission.

The number of phonons whose frequencies lie in a small interval $\Delta\omega$ about ω_s increases in proportion to N. Therefore the

---

* To find the one-phonon transition probability of only one of the system of N oscillators, it is necessary to multiply the probabilities (10.5) and (10.6) by the zero-phonon transition probability for all of the remaining N-1 oscillators (see Exercise 10).

number of one-phonon transitions involving the creation of a single phonon in different oscillators having a frequency ω_s' in the interval

$$\omega_s - \frac{\Delta\omega}{2} < \omega_s' < \omega_s + \frac{\Delta\omega}{2}. \tag{10.7}$$

also increases in proportion to N. Each of the N transitions makes a small contribution, of the order of N^{-1} (infinitesimal when $N \to \infty$),to the intensity of the spectrum in this frequency range. The total contribution of transitions where one phonon is created to this small part of the spectrum has a finite value. This holds for all frequencies ω_s' within the frequency limits for one-phonon transitions, i.e., within the band vibration frequency spectrum.

Furthermore, with large N the band vibration frequencies in the spectrum are distributed so densely that the interval between frequencies is much less than the width of an excited vibrational level. Actually the distance $\delta\omega$ between frequencies can be estimated from the formula $\delta\omega \approx \omega_{max} N^{-1}$, where ω_{max} is the highest band vibration frequency. Therefore the band vibration frequency spectrum can be regarded as continuous in many physical situations.

As a result we conclude that transitions in which a phonon is created lead to the formation of a continuous spectrum around the purely electronic line in a frequency region given by

$$\omega_0 \pm (\omega_{min}^{(r)} + \omega_{max}^{(r)}), \tag{10.8}$$

where $\omega_{min}^{(r)}$ and $\omega_{max}^{(r)}$ denote the lowest and highest frequencies in the r-th band vibration branch. We take the plus sign for absorption and the minus sign for emission. The results are illustrated in Fig. 8.

We stress that we have only explained the part of the spectrum in which the contribution from transitions where one phonon is created is distributed in the form of a continuous background. The distribution of intensity within the region "allowed" for these transitions is determined by the shifts in the equilibrium positions of the band oscillators during the electronic transition. To be more exact it is determined by the dependence of a_s^2 on s and how many oscillators s have a frequency between ω and $\omega + \Delta\omega$. If we regard the phonon frequency spectrum as continuous, then the intensity distribution in the background is determined by the

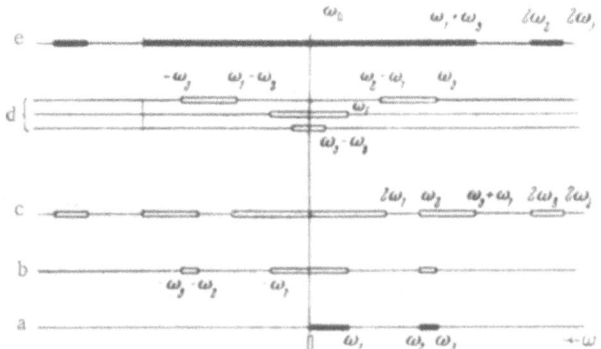

Fig. 8. Regions of the spectrum corresponding to one and
two-phonon transitions in the presence of interactions with
acoustic vibrations $(0 - \omega_1)$ and one optical vibration band
$(\omega_2 - \omega_3)$. a) phonon frequency spectrum; b) one-phonon
transition frequencies; c) two-phonon transitions with creation or annihilation of two phonons; d) two-phonon transitions with the creation of one and annihilation of another
phonon; e) frequency range corresponding (for the ω_1, ω_2,
and ω_3 chosen) to all possible one and two-phonon transitions.

product $a^2(\omega)\rho(\omega)$, where $\rho(\omega)$ is the density of band vibration frequencies and $a^2(\omega)$ is the equilibrium position shift function obtained from a in going to the continuous spectrum $\bar{\omega}_s$.

　　　Figure 9 shows schematically the contribution of one-phonon
transitions to the absorption and emission intensity in the presence of an interaction with three vibrational branches.

　　　Quite often this product has sharp maxima as a function of
ω. Then there are sharp peaks in the vibrational background arising from one-phonon transitions. Our analysis of the contribution
of phonon transitions in this and subsequent sections is also valid
when there are peaks in $a^2(\omega)\rho(\omega)$. However, for simplicity in
discussing and illustrating the nature of the spectrum we limit
ourselves to cases where no peaks appear.

　　　We find a similar qualitative picture for the contribution
from transitions accompanied by the annihilation of one phonon.
The difference lies in the fact that the "allowed" frequency region
shifts to the long-wavelength side of ω_0 in the absorption spectrum

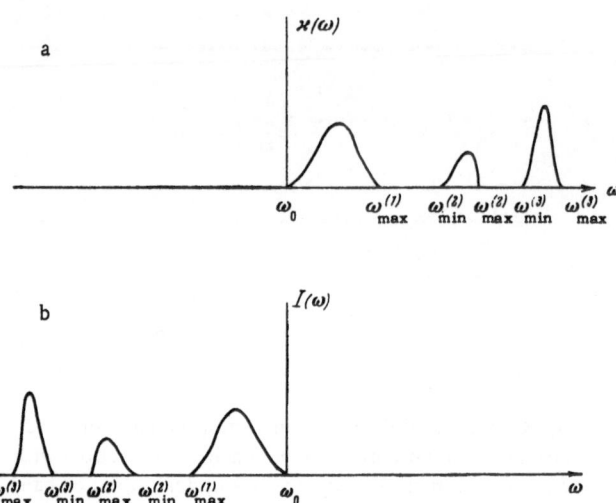

Fig. 9. The purely electronic line at ω_0, and the contributions of transitions in which one phonon is created to the absorption spectrum $\varkappa(\omega)$ (a) and the emission spectrum $I(\omega)$ (b) in the presence of three vibrational branches. The shape of the curve is given by the product $a^2(\omega)\rho(\omega)$, where $a^2(\omega)$ is the shift in the equilibrium position and $\rho(\omega)$ is the density of phonon frequencies.

and to the short-wavelength side in emission. These transitions thus contribute to the anti-Stokes region of the spectrum.

Transitions where a phonon is annihilated are only possible when there are phonons in an excited state before the transition, i.e., the quantum numbers are greater than zero ($i = 1, 2, \ldots$). If we consider a crystal which is in vibrational thermal equilibrium before the transition, then the probability of finding the s-th normal mode in the i-th vibrational state (the probability of finding i_s phonons in the s-th normal mode) is given by the Boltzmann formula.

At absolute zero the occupation probabilities of states with $i_s > 0$ are zero and transitions in which a phonon is annihilated do not occur. When T > 0 the probabilities of excited vibrational states are not zero and it becomes possible to have transitions in

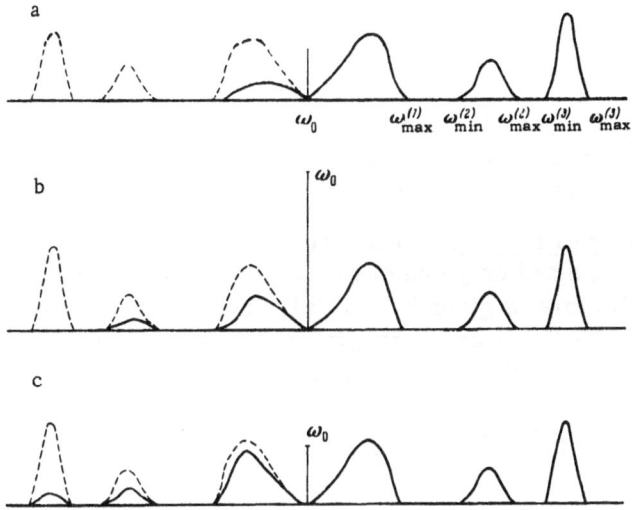

Fig. 10. Contributions of one-phonon transitions to the Stokes and anti-Stokes parts of the spectrum. a) $kT \approx h\bar{\omega}$; b) $kT \approx h\omega_{max}^{(1)}$; c) $kT \approx h\omega_{max}^{(2)}$.

which a phonon is annihilated. At low temperatures this phonon is in one of the low-frequency acoustic vibrations, and with increasing temperature it is possible to have transitions corresponding to annihilation of high-frequency phonons also. As a result there is an additional pronounced temperature dependence* which is characteristic of the anti-Stokes part of the spectrum (Fig. 10).

We again stress that the intensity distribution (within the region accessible to one-phonon transitions) is determined by the squares of the equilibrium position shifts, $q_{s0}^2 = a_s^2 N^{-1}$. The distribution of a_s^2 over the normal modes is a property of the particular impurity center and ultimately determines the vibrational structure of the spectrum. We will return to a more detailed discussion of the role of a_s^2 later (section 15).

---

* As previously noted, according to (10.6) there is an increase in the transition probability with increasing initial excitation level, analogous to Stokes transitions. An additional temperature dependence in the anti-Stokes part of the spectrum arises from the temperature dependence of the occupation of excited vibrational levels, which is much more pronounced in the low-temperature region.

Exercise 8.   Verify equations (10.4a) and (10.4b) using the explicit forms of the harmonic oscillator wave functions (see II.52).

Exercise 9.   Determine the minimum lifetime of a vibrational state where the band vibration spectrum is regarded as quasicontinuous, $k_{max} = 200$ cm^{-1}, and $N = 10^{20}$.

Exercise 10.   Show that the one-phonon transition intensity in one and only one of the oscillators of the system of N band oscillators is given by multiplying by the zero-phonon line intensity I(T) (7.18) and averaging over the temperature factors of $<W(i \to i \pm 1)>$ T (10.5) and (10.6).

§ 11.   Two-Phonon and Multiphonon

Transitions

We now turn to the first class of multiphonon transitions — two-phonon transitions. We will find out whether these transitions make a finite contribution to the spectrum, and if they do, what the nature of the contribution is. The two phonons which participate in a two-phonon transition can either be the same phonon or two different phonons. Therefore, there are in all N^2 possible different two-phonon transitions, corresponding, for example, to the creation of two phonons. It is also possible to have transitions where two phonons are annihilated, and others where one phonon of the s-th mode is created and one of the s'-th mode annihilated (s $\neq$ s').

In order to find the order of magnitude of the line intensity corresponding to a single two-phonon transition we turn to Eqs. (10.2)-(10.6). The two-phonon transition probability involving participation of phonons from different modes is evidently given by the product of the probabilities of the two corresponding one-phonon transitions. For example, for a transition in which one phonon is created in the s-th mode and one phonon in the s'-th mode, we have from (10.5)

$$W\left(i_s, i_{s'} \to i_s + 1, i_{s'} + 1\right) = \frac{1}{4\,h^2 N^2}\,(i_s + 1)\,(i_{s'} + 1)\,a_s^2 a_{s'}^2\,\omega_s \omega_{s'}, \qquad (11.1)$$

where i_s and $i_{s'}$ are the level numbers of the s-th and s'-th oscillators before the transition, and $q_{s0} = a_s N^{-1/2}$ and $q_{s'0} = a_{s'} N^{-1/2}$ are the shifts in the equilibrium positions. It is evident from (10.6) that the intensity is also of the order of N^{-2} for lines corresponding to a two-phonon transition with annihilation of two phonons or annihilation of one phonon and creation of another in a different mode.*

To calculate the transition probability $W(i_s \rightarrow i_s \pm 2)$ when two phonons are created (annihilated) in the same oscillator, we retain in the expansion (10.2) quadratic terms which in this case make a nonzero contribution to higher order quantities. We find $J_{i_s, i_s \pm 2} \sim N^{-1}$ for the overlap integral, and $W(i_s \rightarrow i_s \pm 2) \sim N^{-2}$ for the probability.

Thus each individual two-phonon transition gives a line with a small intensity of the order of N^{-2}. The number of transitions contributing in a small finite frequency range $\Delta\omega$ is proportional to N^2. Consequently, two-phonon transitions contribute a finite intensity to some parts of the continuous spectrum.

It remains to be seen what frequency region these contributions appear in. We let a certain phonon branch be bounded by frequencies ω_{min} and ω_{max}. Then all possible combinations of two phonon frequencies occur in the following regions (reckoned from the purely electronic line): from $\pm 2\omega_{min}$ to $\pm 2\omega_{max}$ (transitions where two phonons are created or annihilated) and from 0 to $\pm(\omega_{max} - \omega_{min})$ (transitions where one phonon is created and one annihilated). If there are several branches we also have transitions in which phonons of different branches participate. In Fig. 8 we show the two-phonon transition frequencies in the presence of interactions with one acoustic and one optical vibration branch. It is clear from this figure that in principle two-phonon transitions can occur over a quite broad region of the spectrum. However, the intensity distribution in fact depends, as for one-phonon transitions, on a_s^2.

We also consider multiphonon transitions in which n phonons (n = 3, 4, 5, ...)participate. Using Eqs. (10.4-10.6) and Eq. (10.2) with the expansion carried to terms q_{s0}^n, we easily see that the n-phonon transition probability contains N^{-n} as a factor.† On the

* The factor I(T) which appears because of the requirement that the states of the remaining N-2 modes do not change, is a finite quantity (see Exercise 10).

† Note that this result is also valid when the frequency of the oscillator changes during the electronic transition [72, 73].

other hand, the number of n-phonon transitions in a small finite frequency range is proportional to N^n. It thus follows that multiphonon transitions make a finite contribution distributed as regions of continuous spectrum about the purely electronic line. These regions include, in principle, all regions allowed as algebraic sums of frequencies in the phonon spectrum of the crystal. In fact, the intensity distribution is determined by the quantities a_s^2 which characterize the particular impurity center and the interaction of the particular transition with the vibrations. It is not difficult to establish that the intensity of a line corresponding to a single n-phonon transition contains (in addition to N^{-n} and the finite factor $I(T)$, see Exercise 10) the product of n factors a_s^2 with the s corresponding to the particular set of modes participating in the n-phonon transition.

§ 12.   The Purely Electronic Line
and the Vibrational Background

We have now shown that the transitions lead to the formation of a continuous vibrational background in the absence of sharp peaks in the product $a^2(\omega)\rho(\omega)$. This background is of multiphonon origin, i.e., it is not generally speaking limited to one-phonon transitions despite the fact that the shift in the equilibrium position of each of the band vibrations becomes infinitesimal in the limit $N \to \infty$.

The width of the background in the absorption or luminescence spectrum cannot be directly compared to the widths of the branches in the vibrational spectrum of the crystal. It may be either substantially larger (due to multiphonon transitions) or substantially less (due to the smallness of a_s^2 for some of the vibrations) than the branch width.

The particular distribution of the background intensity is determined in the simple case which we are considering here by the frequency distribution function and the shifts in the equilibrium positions of the band oscillators. The role of this function (Stokes loss function) is discussed in [40, 74] and examples of it can be found in the experimental spectra given in [75, 76].

There is one special- but important- case: weak interaction between the electronic transition and the vibrations. In our model this corresponds to small a_s.* According to (11.1) the intensity of a two-phonon transition is proportional to $a_s^2 a_{s'}^2$, i.e., to the square of the small quantity a_s^2, while the intensity of a one-phonon transition depends linearly on this quantity. Therefore for very small a, i.e., for small interaction between the transition and the vibrations, the contribution of two-phonon transitions is quite negligible. Since the probabilities of the remaining multiphonon transitions contain a factor a_s^{2n} where n is the number of phonons participating in the process, their contributions can also be neglected with small a_s. We conclude that for very weak interactions the vibrational background about the purely electronic line comes from one-phonon transitions. In this case (and only in this case) can we regard the observed spectrum as arising directly from one-phonon transitions.

Here we must keep in mind that with increasing temperature the contributions of multiphonon processes increase more rapidly than those of one-phonon processes. From Eq. (11.1), using two-phonon transitions as an example, we see that at high temperatures, where $\overline{i}_s$ and $\overline{i}_{s'} \gg 1$ their probabilities grow as the product $i_s i_{s'}$, while the one-phonon transition probability increases as i_s. Therefore it can happen that spectra which are properly interpreted as one-phonon spectra at low temperatures may require multiphonon corrections with increasing temperature.

The integrated intensity of the background increases with temperature.† This already follows from the fact that the integrated intensity of the whole vibronic band is constant in the Condon approximation (section 9) while the integrated intensity of the purely electronic line decreases with temperature (section 7). This conclusion also follows from the phonon transition probabili-

---

* The shift a_s, like q_{s0} in (10.2), will be assumed to be expressed in the dimensionless units used for harmonic oscillators. The smallness of a_s in comparison to unity then means a small shift in the equilibrium position in comparison to $(h/m_s \omega_s)^{1/2}$, i.e., to the "classical zero-point amplitude" of the s-th oscillator. A clearer viewpoint can be presented without introducing normal modes — the Stokes loss is small if the cartesian shifts in the equilibrium positions of the atoms in the impurity center are small in comparison to the average zero-point motion of these atoms.

† This does not ultimately mean that with increasing temperature there is an increase in the intensity at each frequency ω; when the integrated background intensity grows it is accompanied by a broadening arising from the increased contributions of anti-Stokes and multiphonon transitions.

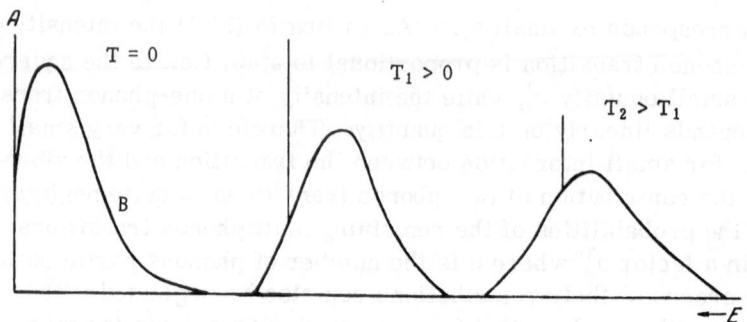

Fig. 11. Temperature dependence of the purely electronic line (A) and the vibrational background (B) (theory for the basic model).

ties (10.2) and (11.1) — with increasing temperature i_s increases, and the transition probabilities increase also.

Since the purely electronic line does not broaden with temperature in the simple case under consideration, we can draw a unique picture of the temperature dependence — the purely electronic line weakens without broadening, and the area "disappearing" from under it, because of the "law of conservation of spectrum area," goes into the integrated background intensity (Figs. 11, 12).

In Fig. 12 we show experimental data [30] on the region of the CdS luminescence spectrum near the line at $\lambda = 4888.6$ Å. We see fair agreement between theory (for the basic model) and experiment. However, the observed half-width of the purely electronic line is about 2 cm^{-1}, i.e., it exceeds the radiation width by three or four orders of magnitude.

The pattern shown in these figures corresponds to an interaction with smooth (in the sense of no sharp maxima of $a^2(\omega)\rho(\omega)$) regions of the acoustic phonon spectrum. Often the structure of the vibrational background of a quasiline spectrum is substantially more complicated. In particular, narrow maxima in the product $a^2(\omega)\rho(\omega)$ can give rise to quite narrow peaks in the vibrational background. This occurs chiefly in the presence of narrow vibrational bands or singularities due to the crowding together of a large number of frequencies at the limit of a vibrational branch if the electronic transition interacts with these branches, or singular

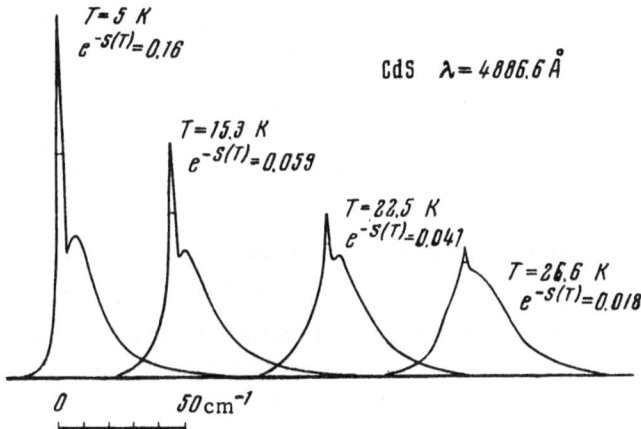

Fig. 12. Experimental data of Gross, Razbirin, and Permogorov
[30] on the temperature dependence of the spectral region of CdS
luminescence near the line with λ = 4888.6 Å. We easily see that
the intensity of the purely electronic line decreases rapidly with
temperature, but the width of this line remains unchanged. However, the experimental half-width of the purely electronic line is
about 2 or 3 cm^{-1}, i.e., it exceeds the radiation width by three or
four orders of magnitude.

regions in the vibration spectrum. In this case the maximum
arises from the factor $\rho(\omega)$. In the second case narrow peaks can
appear in "smooth" regions of $\rho(\omega)$ if the electronic transition interacts selectively and strongly with a narrow range of frequencies
ω. Then the maximum in the product $a^2(\omega)\rho(\omega)$ arises from a maximum in the function $a^2(\omega)$.

In describing narrow maxima in the background it is appropriate to compare them with peaks arising from localized vibrations. Therefore we first consider the effect of localized vibrations.

Exercise 11. 1. Find how many of N oscillators of frequency ω are in excited states at temperature T.

2. Is it correct to assume that the anti-Stokes part of the
spectrum increases with increasing temperature only due to the
decrease in the integrated intensity of the purely electronic line?

3. Calculate the two-phonon transition probabilities $W(i_s \rightarrow i_s \pm 2)$, $W(i_s, i_{s'} \rightarrow i_s - 1, i_{s'} + 1)$, and $W(i_s, i_{s'} \rightarrow i_s - 1, i_{s'} - 1)$.

4. Determine the frequency regions corresponding to one and two-phonon transitions in the presence of one acoustical and two optical vibration branches.

5. Calculate the three-phonon transition probabilities $W(i_s, i_{s'}, i_{s''} \rightarrow i_s + 1, i_{s'} + 1, i_{s''} + 1)$ and $W(i_s \rightarrow i_s + 3)$.

6. Show that the n-phonon transition probability in a system of N phonons ($N \gg n$) is proportional to $I(T)N^{-n}$ where $I(T)$ is the intensity of the purely electronic line.

§ 13.   The Effect of Localized Vibrations

We add one localized vibration having frequency Ω to the model. The potential curve of the localized vibration changes during the electronic transition by a finite, sometimes quite large, amount.* In particular, the shift in the equilibrium position (in dimensionless units) is finite. Therefore the method used to calculate the overlap integrals for the band vibrations, based on expansion of the wave function in the small parameter $q_{s0} = a_s N^{-1/2}$, is not applicable. The overlap integrals must be calculated explicitly. In our case (neglecting changes in the interatomic force constants) this is comparatively simple.

As a further simplification we will assume that the temperature of the crystal is low enough that thermal excitation of the localized vibrations can be neglected. This is equivalent to the assumption that $\hbar\Omega \gg kT$. In this case one need not average over initial localized vibrational states; all transitions in which the localized vibration participates begin in its ground state.

The assumption $\hbar\Omega \gg kT$ is often a good approximation since localized vibration frequencies are quite high — the quantum energy $\hbar\Omega$ is usually from 0.01 to 0.1 eV. Therefore always at the temperatures of liquid helium ($T \approx 4$ K), and often at the temperature of liquid nitrogen ($T \approx 77$ K), this condition is satisfied. At the end of the section we give (without derivation) the results

---

* The potential curve for a localized vibration and the changes in its parameters agree generally with the potential curves for diatomic molecules (see, for example, [3]).

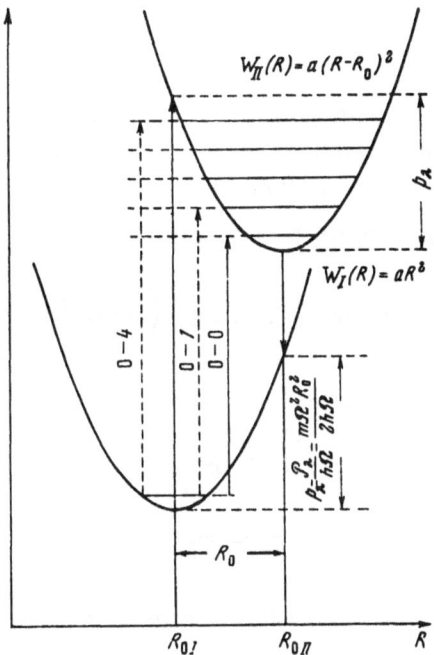

Fig. 13. Calculation of the Franck—Condon
integrals for a localized vibration.

obtained taking thermal excitation of the localized vibration into
account.

The (acoustic) vibration frequency spectrum begins at zero,
and thermal excitation of acoustic vibrations must always be taken
into account. For long-wavelength acoustic vibrations even helium
temperatures are high. The role of the band vibrations in the formation of the spectrum in the presence of localized phonons will
be discussed somewhat later.

We now proceed to a calculation of the overlap integrals for
transitions from the zeroth vibrational level of a harmonic oscillator in which only the equilibrium position changes as a result of
the electronic transition (Fig. 13). The material given will also
supplement the quantum-mechanical Franck—Condon principle.

The harmonic oscillator wave function, $\varphi_f(\xi)$, takes the form
of a Hermite polynomial multiplied by a gaussian weight function
and can be written as (see Appendix II)

$$\varphi_f(\xi) = \frac{(-1)^f}{\sqrt{2^f f! \sqrt{\pi}}} e^{\xi^2/2} \frac{d^f}{d\xi^f} e^{-\xi^2}. \tag{13.1}$$

Here f is the number of the energy level and ξ is the distance expressed in the dimensionless units used for a harmonic oscillator, $\xi = R/\bar{R}$, where R is the distance (in cm) and $\bar{R} \equiv (h/m\Omega)^{1/2}$.

The overlap integral takes the form

$$J_{of} = \int_{-\infty}^{+\infty} \varphi_{IIf}(\xi)\,\varphi_{I0}(\xi)\,d\xi = C_f \int_{-\infty}^{+\infty} e^{-\xi\xi_0} \frac{d^f}{d\xi^f} e^{-\xi^2} d\xi. \tag{13.2}$$

for $0 \rightarrow f$ transitions ($f \rightarrow 0, 1, 2, \dots$). Here we take the origin at the equilibrium position of the oscillator in electronic state II, so that $\varphi_{I0}(\xi) = \varphi_{II0}(\xi + \xi_0)$, and introduce the notation

$$C_f \equiv \frac{(-1)^f}{\sqrt{2^f f! \sqrt{\pi}}} e^{-\xi_0^2/2}.$$

The integral in (13.2) can be transformed, by integrating by parts f times, to

$$J_{of} = C_f \xi_0^f \int_{-\infty}^{+\infty} e^{-\xi^2 - \xi\xi_0} d\xi = C_f \xi_0^f e^{\xi_0^2/4} \int_{-\infty}^{+\infty} e^{-\left(\frac{\xi_0}{2} + \xi\right)^2} d\xi = \frac{(-\xi_0)^f e^{-\xi_0^2/4}}{\sqrt{2^f f!}} \tag{13.3}$$

Thus we have for the probability of interest

$$W_{of} = J_{of}^2 = e^{-\xi_0^2/2} \left(\frac{\xi_0^2}{2}\right)^f \frac{1}{f!} = e^{-p_\lambda} \frac{p_\lambda^f}{f!}. \tag{13.4}$$

Here the Stokes loss for the localized vibration is $p_\lambda \equiv \xi_0^2/2$, or, if the shift in the equilibrium position is put in dimensional units, $p_\lambda = m\Omega^2 R_0^2/2h\Omega$ (see section 7).

In Fig. 14 we show three typical patterns for the probabilities of transitions from the zeroth level of the localized oscillator corresponding to: a) small ($p_\lambda = 0.1$); b) medium ($p_\lambda = 1$); and c) large ($p_\lambda = 10$) Stokes losses, calculated from (13.4). The patterns correspond to the intensity distribution in the low–temperature ($kT \ll h\Omega$) spectrum of a crystal containing impurities where there is no interaction of the electronic transition with band vibrations. We also neglect anharmonic interactions between localized and band vibrations, which lead to a broadening of the excited states of

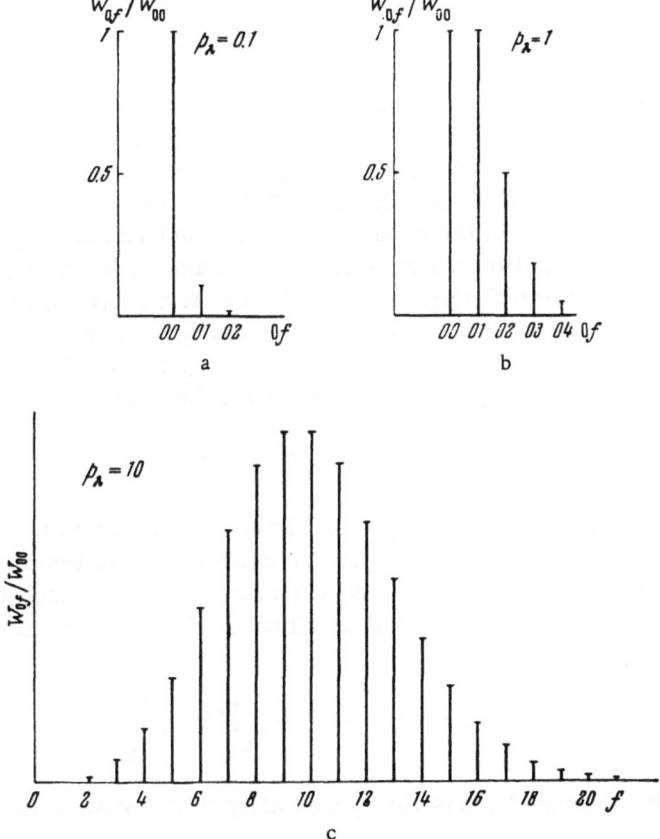

Fig. 14. Three typical distributions of the vibronic transition probability. Along the ordinate we show the ratio of the probability W_{0f} to the probability W_{00} of the $0-0$ transition. In cases a) and b) $W_{00} = 1$, while in case c) the units are arbitrary.

the localized vibration. The patterns shown in Fig. 14 also correspond to the intensity distribution among vibrational bands in the spectrum of a diatomic molecule (neglecting internal structure of the band arising from rotation and translation of the molecule and the effect of vibrational-rotational coupling on the distribution of intensity between vibrational bands).

We now consider the fact that an impurity center in a crystal always has interactions with the band vibrations. In our model each normal mode is completely independent of the others. It follows that the probability of a transition in one normal mode does not depend on what transitions in the other modes accompany the electronic transition. Therefore when an arbitrary number of quanta are created or annihilated in arbitrary (localized or band) vibrations the transition probability is a product of creation (anni-hilation) probabilities for the corresponding number of quanta in the individual oscillators. Equation (7.5) (which is also valid, as we can easily see, in the presence of localized vibrations) ex-presses this situation mathematically. Actually, for a single localized vibration the probability of a I → II electronic transition in which the vibrational state of the crystal goes from an initial state Iλi to a final state II$\lambda'f$ can, according to Eq. (7.5), be writ-ten in the form (the electronic indices will be dropped subsequently)

$$W\left(\text{I}\lambda i_s \rightarrow \text{II}\lambda' f_s\right) = W_{\lambda\lambda'} \prod_s W_{i_s f_s}. \tag{13.5}$$

Here λ denotes the state of the localized vibration and i_s denotes the state of the whole set of other vibrations. If the temperature is too low to excite the localized vibration ($kT \ll h\Omega$) then the probability, averaged over initial vibrational states in thermal equilibrium, is

$$\langle W\left(0 i_s \rightarrow \lambda' f_s\right)\rangle_T = W_{0\lambda'} \left\langle \prod_s W_{i_s f_s}\right\rangle_T, \tag{13.6}$$

where the averaged transition probability in the system of band vibrations $\left\langle \prod_s W_{i_s f_s}\right\rangle_T$, with a precision as in the absence of local-ized vibrations (compare with Eq. (7.5)), while $W_{0\lambda'}$ is given by (13.4).

We first consider the integrated intensity of the transitions corresponding to a single definite $0 \rightarrow \lambda'$ transition of the localized vibration. To obtain the corresponding probabilities it is neces-sary to average over initial states and sum over all possible final states of the band vibration in (13.6). We then have

$$\sum_{[i_s]} n_{i_s} \sum_{[f_s]} W\left(0 i_s \rightarrow \lambda' f_s\right) = W_{0\lambda'} \sum_i n_i \sum_f W_{if} \equiv W_{0\lambda'} S_{0\varkappa}(T). \tag{13.7}$$

In the Condon approximation the zeroth moment with respect to the band vibration is $S_{0\varkappa}(T) = S_{0\varkappa} = \text{const}$.

Here we transform to new simplified summation indices by making the substitutions $[i_s] \to i$ and $[f_s] \to f$. The corresponding sums over i and f include summation over all possible initial i and final f of the band vibrations. Since the wave functions for each band oscillator form a complete set for the corresponding normal mode, all of the equations of the moments method remain valid (see Appendix VI), particularly the conclusions given in section 9 about the integrated intensity of the whole vibronic band. Therefore we can legitimately use these results, according to which in the Condon approximation the zeroth moment $S_{0\varkappa}(T)$ of the distribution is constant.

We can now formulate the first result for our model: the integrated intensities of the bands which correspond to the $0 \to \lambda'$ transitions in the localized vibration are given by the squares of the Franck–Condon integrals $W_{0\varkappa}$ for the localized vibration according to (13.4). The interaction with the band vibrations does not affect the distribution of the integrated intensity in the series of localized vibration bands.

The band vibrations evidently give rise to the internal vibrational structure of each band. It is not difficult to see from (13.6) that this structure is the same in each band. Actually, the ratio of the probabilities of two transitions $0 \to \lambda'$ and $0 \to \lambda''$ in which the same set of band vibration quanta participate is

$$\frac{\langle W(0\,i \to \lambda'f)\rangle_T}{\langle W(0\,i \to \lambda''f)\rangle_T} = \frac{W_{0\lambda'}}{W_{0\lambda''}} \frac{\left\langle \prod_s W_{i_s f_s}\right\rangle_T}{\left\langle \prod_s W_{i_s f_s}\right\rangle_T} = \frac{W_{0\lambda'}}{W_{0\lambda''}}, \qquad (13.8)$$

i.e., it is determined by the ratio of the corresponding transition probabilities of the localized vibration. The first of the transitions contributes to the spectrum with a frequency ω_1 shifted from the purely electronic line by a distance $\lambda'\Omega + \Sigma_s \omega_s (f_s - i_s)$ and the second of them is at a frequency $\omega_2 = \lambda''\Omega + \Sigma_s \omega_s (f_s - i_s)$. This holds for all transitions involving band vibrations. It follows from this that there is a "similarity law" for the internal structure of all the bands in the series arising from localized vibrations (Fig. 15).*

---

* The conclusions about the integrated intensity obtained above (Eq. 13.7) also follow from the "similarity law."

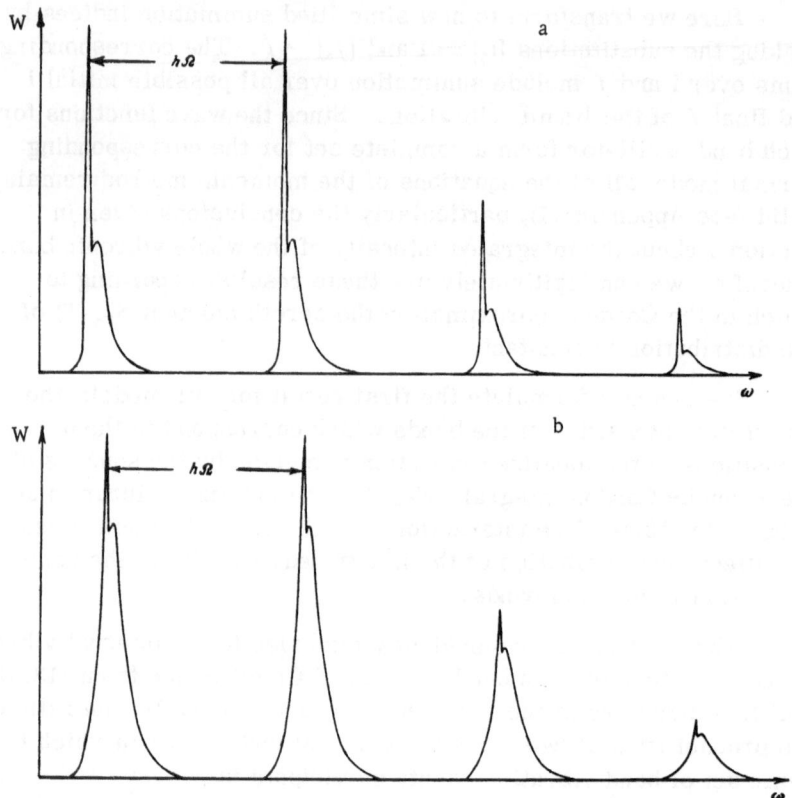

Fig. 15. Electronic transition interacting with a localized vibration of frequency Ω and with band vibrations (a) 5 K; b) 22.5 K). The Stokes loss for the localized vibration is taken as $p'_\lambda = 1$ (Fig. 14b); the interaction with the band vibrations is taken from Fig. 12.

It remains to explain what kind of internal structure the bands have. On the strength of the "similarity law" it suffices to consider any one of them. The structure is determined by the dependence of $\langle \Pi W_{i_s f_s} \rangle_T$ on the frequency of the transitions $i_s \rightarrow f_s$ in Eq. (13.6) with any fixed indices 0 and λ. We know that it is precisely the same as the vibrational structure of the vibronic band in the absence of a localized vibration. In particular, in each of the bands there is a replica of the purely electronic line located at a frequency $\lambda'\Omega$ for the $0 \rightarrow \lambda'$ band. The vibrational background arises in general from multiphonon transitions in band vibrations. The temperature dependences of these replicas of the purely electronic line and the vibrational background are exactly as in the spectrum in the absence of a localized vibration: with increasing temperature the intensity of the replica of the purely electronic line decreases according to Eqs. (7.17a) or (7.18), and the intensity which disappears from it, due to an "area conservation law," is added to the integrated background intensity (Fig. 15). The explicit dependence of the background intensity is given by $a^2(\omega)\rho(\omega)$ for the band vibrations.

An overall view of the role of the localized vibration can be formulated as follows. The interactions of the electronic transition and the band vibrations with a localized vibration lead to the appearance of a series of replicas of the vibronic spectrum which would appear if the system interacted in the same way with the band vibrations* in the absence of any localized vibration. These replicas are exactly similar to the spectrum in the absence of a localized vibration and are located at distances $\pm\lambda'\Omega$ ($\lambda' = 0, 1, 2, \dots$) from it, with the plus sign corresponding to absorption and the minus sign to luminescence. The intensity of the band is given by the factor $W_{0\lambda'}$ in Eq. (13.4).

In other words, if the interaction with the band vibrations, in which the vibronic band has the shape $I_\varkappa(\omega, T)$ (basic shape), is supplemented by an interaction with one localized vibration then the spectrum breaks up into a sum of replicas of the basic shape

$$I(\omega, T) = W_{00}I_\varkappa(\omega, T) + W_{01}I_\varkappa(\omega \mp \Omega, T) + W_{02}I_\varkappa$$

$$\times (\omega \mp 2\Omega, T) + \dots + W_{0\lambda'}I_\varkappa(\omega \mp \lambda'\Omega, T) + \dots \qquad (13.9)$$

The minus sign corresponds to absorption and the plus sign to luminescence.† Two or several terms can make comparable contributions to the total intensity at a frequency ω.

The assumption of low thermal excitation of the localized vibration was made for simplicity. It is not difficult to see that when thermal excitation of the localized vibration is taken into account the results formulated above change in two respects. First, replicas $I_\varkappa(\omega, T)$ appear in the anti-Stokes region of the spectrum.

---

* In the sense of the function $a^2(\omega)\rho(\omega)$.

† The basic shapes for absorption and emission are different; they are mirror images of each other relative to the purely electronic line at ω_0.

In the second place the factor $W_{0\lambda}$ is replaced by a thermally averaged quantity (see (13.10) below).

We further note that if the results obtained are to be valid, as we easily see from their derivation, it is not necessary that the vibrations be harmonic or that each normal mode be completely independent. It suffices that the localized mode be independent of the band modes. Anharmonic interactions between the band vibrations or changes of their axes during the electronic transition do not affect the similarity law of the internal vibrational structure of the band, even though the structure itself and its temperature dependence may be substantially different from the basic version of the theory. Anharmonicity or changes in the localized vibration frequency do not lead to any violation of the similarity law only if there is no interaction between localized and band vibrations. Ultimately there is a change in the Franck—Condon factors and the energy spectrum of the localized mode.

We now give without derivation a formula for the temperature dependence of the integrated intensity distribution (zeroth moment) in the series of quasilines which are replicas of the purely electronic line arising from a single localized vibration:

$$S_0^k \, (T) = F_\lambda \, (T) \sum_\lambda \exp\left[- \hbar\Omega \left(\lambda + \frac{1}{2} \right) / kT \right] W_{\lambda, \lambda + k} I \, (T)$$

$$= \exp[-p_\lambda(2\bar{n}_\lambda + 1)](1 + \bar{n}_\lambda^{-1})^{k/2} I_k(2p_\lambda \sqrt{\bar{n}_\lambda \, (\bar{n}_\lambda + 1)}) \, I(T). \quad (13.10)$$

Here k is the number of the replica of the basic shape, λ is the localized mode index, $\bar{n}_\lambda = [\exp(-\hbar\Omega/kT) - 1]^{-1}$ is the average occupation number of the λ-th mode in thermal equilibrium, and $I_k(x)$ is the k-th order modified Bessel function (see Appendix III). Equation (13.10) can be represented as two factors:

$$S_0 \, (T)_k = S_{0\lambda} \, (T)_k I \, (T), \quad (13.11)$$

where I(T) is the temperature dependence of the integrated intensity of the purely electronic line in the absence of a localized vibration, as given by (7.17a), and $S_{0\lambda} \, (T)_k$ gives the localized vibration contribution. More precisely, I(T) is the integrated intensity of the purely electronic line in the basic shape, with the condition that the whole integrated intensity under the basic shape (i.e., the purely electronic line and its wings) is normalized to unity. $S_{0\lambda} \, (T)_k$

is a weight factor by which the k-th replica of the basic shape is multiplied. The intensity of the vibrational wing at a distance ω from the quasiline in the k-th replica of the basic shape is obtained from (13.10) if the intensity $S_0(T)_k$ of the quasiline is multiplied by the ratio $I_x(\omega, T)/I(T)$ where $I_x(\omega, T)$ is the normalized intensity at a distance ω from the purely electronic line in the basic shape. With $\hbar\Omega \gg kT$, $S_{0\lambda}(T)_k$ becomes W_{0k} (Eq. (13.8), where we have taken $\lambda' = k$).

We have for the integrated intensity of the purely electronic line in the presence of a localized vibration

$$S_0^k(T)_{k=0} = S_0^0(T) = \exp[-p_\lambda(2\bar{n}_\lambda + 1)]$$
$$\times I_0(p_\lambda[\text{sh}\,(\hbar\Omega/2\,kT)]^{-1})\exp\left[-\sum_x p_x(2\bar{n}_x + 1)\right] = S_{0\lambda}(T)_0 I(T), \quad (13.12)$$

where $I(T)$ and $S_{0\lambda}(T)_0$ are factors arising from the band and localized vibrations, respectively, and $x = (i/2)(e^x - e^{-x})$ is the hyperbolic sine.

We now turn our attention to a curious fact: the factor $S_{0\lambda}(T)_0$ can have a nonmonotonic temperature dependence. This happens if the localized vibration has a substantial Stokes loss and can (for a small total Stokes loss of the band vibrations) cause the integrated intensity of the purely electronic line to increase initially when the temperature is raised above absolute zero (in the temperature range where $kT \approx \hbar\Omega$) and only then (with $kT \gg \hbar\Omega$) begin to drop off. This result is easily interpreted in terms of the semiclassical Franck—Condon principle as a consequence of thermal broadening of the localized vibration coordinate distribution in the initial electronic state (see Appendix I).

Exercise 12. Calculate the total transition probability $\sum_{\lambda'=0}^{\infty} W_{0\lambda'}$. Discuss the result from the point of view of the "conservation law" for the area under the vibronic band.

Exercise 13. Calculate $S_0^k(T)$ (13.10) using Eq. (III.1) (Appendix III).

Exercise 14. Show the validity of Eq. (13.7) (use the completeness relation for the system of wave functions of each phonon).

Exercise 15. How do the results of section 13 change if we introduce a deviation from the Condon approximation?

Exercise 16. Show that when the Stokes loss p_λ goes to zero, (13.12) goes over to the formula for the integrated intensity of the purely electronic line in the absence of a localized vibration.

Exercise 17. Show that according to (13.12) $S_0^0(T)$ can have a nonmonotonic temperature dependence. Illustrate the result using a model with suitable parameter values as an example.

Exercise 18. Explain the nonmonotonic temperature dependence of the factor $S_{0\lambda}(T)_0$ using the potential curves and the semiclassical Franck–Condon principle.

Exercise 19. Generalize Eq. (13.9) to an anharmonic localized vibration independent of the band vibrations, whose potential corresponds to a Morse function.

§ 14. Several Localized Vibrations

In the harmonic approximation and with no change in the interatomic force constants during an electronic transition where all of the normal modes are independent, the results of the previous section can be trivially generalized to several localized vibrations. The similarity law persists, and receives a more general formulation: with L localized vibrations with frequencies Ω_1, Ω_2, ..., Ω_L the probability distribution function is the sum of the basic shape $I_\varkappa(\omega, T)$ and replicas shifted from it by distances $\lambda_1\Omega_1 + \lambda_2\Omega_2 + \ldots + \lambda_L\Omega_L$, where $\lambda_1, \lambda_2, \ldots, \lambda_L$ are integers 0, ±1, ±2, ... with the plus sign corresponding to absorption and the minus sign to emission. The anti–Stokes replicas of the basic shape are obtained if we associate the annihilation of a quantum of the localized vibration frequency Ω_λ with the negative quantity $-\Omega_\lambda$.

If the temperature is low enough that none of the localized vibrations is excited, the contributions from the anti–Stokes replicas of the basic shape can be neglected and it is also unnecessary to average over the initial localized vibrational states. In this case the total probability distribution is given by

$$I(\omega, T) = \sum_{[\lambda_s]} I_\varkappa \left(\omega \mp \sum_s \lambda_s\Omega_s; T \right) \prod_{\{\lambda_s\}} W_{0\lambda_s}^{(s)} =$$

$$= W_{00}^{(1)}W_{00}^{(2)} \dots W_{00}^{(L)}I_\varkappa(\omega, T) + W_{01}^{(1)}W_{00}^{(2)} \dots W_{00}^{(L)}$$

$$\times I_\varkappa(\omega \mp \Omega_1; T) + W_{00}^{(1)}W_{01}^{(2)} \dots W_{00}^{(L)}I_\varkappa(\omega \mp \Omega_2; T) + \dots$$

$$\dots + W_{01}^{(1)}W_{01}^{(2)}W_{00}^{(3)} \dots W_{00}^{(L)}I_\varkappa(\omega \mp (\Omega_1 + \Omega_2); T) + \dots$$

$$\dots + W_{0\lambda_1}^{(1)}W_{0\lambda_2}^{(2)} \dots W_{0\lambda_L}^{(L)}I_\varkappa(\omega \mp (\lambda_1\Omega_1 + \lambda_2\Omega_2 + \dots + \lambda_L\Omega_L)) \quad (14.1)$$

Here the sum runs over all possible sets $[\lambda_s]$ of changes in the localized vibrational states. For example, the first three terms written above correspond to the sets $\{\lambda_s\} = \{0, 0, \dots, 0\}$, $\{\lambda_s'\} = \{1, 0, 0, \dots, 0\}$, and $\{\lambda''\} = \{0, 1, 0, \dots, 0\}$. The Franck–Condon factor given by Eq. (13.4), i.e., the integrated intensity of the replica of the basic shape which accompanies a transition in which the set $\{\lambda_s\} = \{\lambda_1, \lambda_2, \dots, \lambda_L\}$ is created, is given by

$$\prod_{\{\lambda_s\}} W_{0\lambda_s}^{(s)} = \prod_{\{\lambda_s\}} e^{-p_s}\frac{p_s^{\lambda_s}}{\lambda_s!} = e^{-(p_1 + p_2 + \dots + p_L)}\frac{p_1^{\lambda_1}p_2^{\lambda_2} \dots p_L^{\lambda_L}}{\lambda_1!\lambda_2! \dots \lambda_L!} \quad (14.2)$$

where p_s is the dimensionless Stokes loss in the s-th localized vibration and λ_s is the quantum number of the final state of the s-th localized vibration for the band considered.

The integrated intensities of the quasilines which are replicas of the purely electronic line are (at absolute zero)

$$S_0^{\lambda_s}(T) = \exp\left(-\sum_\varkappa p_\varkappa\right) \prod_{\{\lambda_s\}} W_{0\lambda_s}^{(s)}. \quad (14.3)$$

The intensity at some point on the wing of the $\{\lambda_s\}$-replica is

$$I(\omega, 0) = \prod_{\{\lambda_s\}} W_{0\lambda_s}^{(s)}I_\varkappa\left(\omega \mp \sum_s \lambda_s\Omega_s; 0\right). \quad (14.4)$$

As mentioned previously, even for a single localized vibration one expects comparable contributions to a given part of the spectrum from the various terms in the sum in (13.9). One is even more likely to have a superposition of different replicas of the basic shape in the presence of several localized vibrations. In the part of the spectrum where superposition occurs, the vibrational structure can be so strongly changed that it becomes impossible to recognize the features of the basic shape $I_\varkappa(\omega, T)$. At the same time, in other parts of the same vibronic spectrum the structure of $I_\varkappa(\omega, T)$ may be fairly distinct. With increasing temperature $I_{\varkappa}(\omega, T)$ broadens and its structure becomes smoother.* Moreover, anti-Stokes components of the localized vibrations appear which in turn leads to an increase in the superposition of different replicas of the basic shape. It is thus evident that in general the vibrational structure of the spectrum and its temperature dependence can be quite complicated, despite the fact that the very simple relations (13.9) and (1.41) of replication of the basic shape according to the similarity law are the basis of the phenomena. In interpreting an observed spectrum we must, finally, keep in mind a number of features not considered in the basic model of an impurity center: the possibility of electronic level splitting, the fine details of the interaction of the electronic transition with the vibrations, the presence of several kinds of impurity center, etc. (see Chapter IV). But the basic model, as already noted, is a very well-founded approximation to reality in the theory considered here which is suitable for use in interpreting the vibrational structure of spectra.

In conclusion we note that the formulas for the temperature-dependent intensities of the replicas of the basic shape for L localized vibrations are obtained from Eqs. (13.10)-(13.12) if the factors $S_{0\lambda}(T)_k$ contained in them, which express the effect of a single localized vibration, are replaced by a product of L factors $S_{0s}(T)_{\lambda_s}$ each of which expresses the role of one of the localized vibrations.

For example, we find that we have for the integrated intensity of the purely electronic line with L localized vibrations having frequencies Ω_λ and Stokes losses p_s (s = 1, 2, ... , L)

$$S_0^o(T) = \exp\left[-\sum_\varkappa 2p_\varkappa \frac{k\tau_\varkappa}{\hbar\omega_\varkappa}\right] \exp\left[-\sum_{s=1}^{L} 2p_s \frac{k\tau_s}{\hbar\Omega_s}\right]$$

$$\times \prod_{s=1} I_0\left(p_s \left[\sinh\left(\hbar\Omega_s/2\,kT\right)\right]^{-1}\right) = I(T)\,S_{01}(T)_0 S_{02}(T)_0 \cdots S_{0L}(T)_0.$$

$$\tag{14.5}$$

Exercise 20. Calculate the total integrated intensity in the series of bands which correspond to all possible transitions in one of the localized oscillators, using Eq. (14.2).

---

* In particular, according to (7.18) the intensities of the zero-phonon line and its replicas at the localized frequencies decrease. We will see later that quasilines are also broadened with increasing temperature.

§ 15. Structure of the Vibrational Background. Role of Pseudolocalized Vibrations and Singularities in the Frequency Spectrum of the Band Vibrations

As previously noted, the vibrational background of the quasiline spectrum can have structure. It can even have sharp maxima. This occurs in the presence of a narrow vibration band.

This is demonstrated by investigating the corresponding Franck—Condon factor (sum of the Franck—Condon factors corresponding to vibrations of the same frequency, taking all possible multiphonon processes into account). But it can also be understood from physical consideration of the idea of a vibrational band of zero width.

Actually it follows from the theory of the normal modes of a crystal (Appendix II) that band vibrations can have the same frequency only when the vibrational motion in the unit cell is completely independent of the motion in the rest of the crystal. This kind of band vibration could be regarded as a set of localized vibrations each of which is localized in one unit cell. In an electronic transition in a localized region of the crystal, the interaction with the band vibrations would appear to be interaction with a localized vibration (if the transition affects only one unit cell) or with several localized vibrations of the same frequency (if the electronic transition changes the equilibrium positions in several unit cells). In the latter case the structure is just as in the interaction with one localized vibration but the Stokes loss is the total loss in the localized vibrations.* In other words the interaction of an electronic transition with band vibrations of this kind can be regarded as an interaction with a localized vibration at an impurity center, since a finite shift in the equilibrium position enters in the formulas for the Franck—Condon factors.

---

* Here we refer to the fine details connected with the fact that with a change in the electronic state the conditions of the vibrations change in a localized region of the crystal, i.e., there is a change in the interatomic force constants. Therefore in an excited localized electronic state not all of the unit cells are equivalent for vibrational motion and, strictly speaking, we cannot assume that all of the vibrations in the band have the same frequency as before.

Evidently the above discussion also provides a qualitative interpretation of the situation when a vibrational band has a small but finite width $\Delta\omega (\Delta\omega \ll \bar{\omega})$, where $\bar{\omega}$ is the average frequency in the band. This discussion is applicable also when there are regions with a high density of frequencies in a comparatively wide band (the condition $\Delta\omega \ll \bar{\omega}$ is fulfilled for a certain part of the band).

For a more precise explanation we turn to a treatment of a_s^2 as a function of the band mode number s. It follows from (10.5) that the distribution of the one-phonon transition intensity at absolute zero (i = 0) is proportional to this function. It is quite evident that peaks in the total (multiphonon) vibrational background can appear only when they occur in the one-phonon background at absolute zero. Actually, we see from the discussion of section 11 that the anti-Stokes and multiphonon transitions reproduce the region of the spectrum allowed for one-phonon transitions and carry it into different combinations. The superposition of the vibrational background arising from multiphonon transitions of various orders or increased temperature can lead to reproduction of the peaks occurring in the one-phonon spectrum at absolute zero in other parts of the spectrum. Here the peaks can broaden and their relative intensities decrease in relation to the structureless background but no new peaks which do not appear in the one-phonon background can be formed.

Therefore we focus our attention on the vibrational structure of the one-phonon contribution to the spectrum at absolute zero, which is determined by a_s^2 as a function of the vibration frequency ω. Since the band vibration frequency spectrum is practically continuous it is convenient to go over to a function of a continuous argument, $\omega a^2 (\omega)$, which gives the squares of the shifts in the equilibrium positions of the normal modes during the electronic transition. The function $a^2(\omega)$ is not necessarily continuous. However, within one vibrational band it should be regarded as continuous and smooth.

In going to a continuous spectrum it is necessary as always to consider the frequency density $\rho(\omega)$ of the crystal, i.e., we must take account of the fact that in the same frequency interval $\Delta\omega$ there are different numbers of normal modes s in different parts of the spectrum. As a result we have a function of a continuous

argument which gives the distribution of intensity in the one-phonon Stokes wing in the form*

$$p(\omega)\,d\omega = \frac{1}{2}a^2(\omega)\,\rho(\omega)\,d\omega. \qquad (15.1)$$

The function $p(\omega) \equiv \frac{1}{2}a^2(\omega)\rho(\omega)$ is called the Stokes loss
function [40, 76] since $a^2(\omega)/2N$ is none other than the dimensionless Stokes loss (7.19b) per oscillator in the frequency range from
ω to $\omega + d\omega$, and after multiplying it by the number of oscillators
$N\rho(\omega)$ in this interval we obtain the total dimensionless Stokes loss
in this interval.

It is important for the structure of the vibrational background that sharp maxima of the function $p(\omega)$, which is a product
of two functions according to (15.1), can come either from the frequency density $\rho(\omega)$ or the displacement function $a^2(\omega)$. The sharp
maxima in $\rho(\omega)$ arise from narrow bands or regions where there
is an "accumulation of frequencies" in the vibration spectrum or
from localized vibrations.[†] Sharp maxima appear in $a^2(\omega)$ if the
electronic transition selectively interacts with vibrations in one
narrow frequency range, i.e., if it selectively changes the equilibrium positions of the normal modes in a narrow frequency range,
with the equilibrium positions of the remaining modes nearly unchanged. In this case sharp maxima of $p(\omega)$ can arise in frequency
regions in which the frequency density of the host crystal has no
maxima or singularities. In other words, sharp maxima in the
vibrational background of the vibronic spectrum can also arise
with broad vibrational bands in the host crystal.

The appearance of maxima which are absent in the vibration
spectrum of the host crystal is closely connected with pseudolocalized vibrations (see section 3 and Appendix II). Actually the func-
tion $p(\omega) = \frac{1}{2}a^2(\omega)\rho(\omega)$ can be regarded as the frequency distribution of the potential energy connected with a given change of the

---

* Here we give a (somewhat simplified) equation for one vibrational band. For details of the definition of $p(\omega)$ see [62].

† Since the frequency spectrum of the band vibrations is changed to an infinitesimal
extent by the introduction of an impurity (see Appendix II), narrow bands and regions of "frequency accumulation" (see [77, 78]) should occur just as in the vibrational spectrum of the host lattice. When impurities are introduced a significant
change in $\rho(\omega)$ can occur as a localized vibration. It can be described by a term
$\delta(\omega - \Omega)$ in the frequency density function $\rho(\omega)$.

cartesian equilibrium positions* or, in other words, as the frequency distribution of the energy in a wave packet of vibrations. If this packet now contains a narrow peak, it means that the changes in the cartesian equilibrium positions brought about by the electronic transition correspond to excitation (in addition to the usual excitation of vibrations) of a long-lived vibrational wave packet, i.e., a pseudolocalized vibration. This means that if there are quite sharp peaks in the vibrational background which cannot be explained as localized vibrations or peaks in the vibrational frequency density $\rho(\omega)$, these peaks can be regarded as arising from pseudolocalized (quasilocalized) vibrations.

If there are no peaks in the vibrational background it still does not follow that there is no possibility of exciting pseudolocalized vibrations in a given impurity center. In addition to the trivial reasons which inhibit observations of vibrational structure (superposition of multiphonon transitions, thermal broadening of the quasilines, etc.) one must bear in mind that the electronic transition may simply not interact with the pseudolocalized vibrations possible for the center, just as it need not interact with localized vibrations. This happens if the configuration of cartesian displacements arising from the electronic transition does not correspond to excitation of a localized (pseudolocalized) vibration. Often such a situation arises for symmetry reasons, which also give the corresponding selection rules [79, 80]† for localized vibrations. These selection rules, as with many other conclusions for localized vibrations, can be transferred to pseudolocalized vibrations with results which are more precise for more sharply expressed pseudolocalized vibrations, i.e., ones corresponding to narrow wave packets. In particular, the intensity distribution in a series of peaks arising from well-developed pseudolocalized vibrations can be fairly well described by the equations of section 13, derived for localized vibrations. (Somewhat better equations,

---

* For simplicity we discuss the situation using a classical description of the vibrations. In quantum mechanics one speaks of the potential energy operator and its average values.

† For other processes which occur at an impurity center the selection rules are different and may be more favorable for the appearance of vibrational structure. Thus the impurity Raman scattering or infrared absorption spectra may have peaks from localized or pseudolocalized vibrations which do not appear in the vibronic absorption spectrum.

(15.2)-(15.5) are given below.) Thus the idea of a pseudolocalized vibration as the analog of a localized vibration is useful in the theory of vibronic transitions.

We must not forget, however, that the vibrational wave packet always involves the whole vibration band. In describing a pseudolocalized vibration peak at $\bar{\omega}$ in $p(\omega)$ as a localized vibration of frequency Ω, one must cut off these wings and pay attention to the corresponding part of the interaction with the "true band vibrations." It is clear that the procedure for cutting off the wings is to some degree arbitrary.

It is not arbitrary for a true localized vibration — which corresponds uniquely to a δ-like term for $p(\omega)$, originating from the function $p(\omega)$. Therefore in the h a r m o n i c description of vibrations there is always a difference in principle between localized and pseudolocalized vibrations in which the finite phonon lifetime (due to the interaction between normal modes) is taken into account, this difference persists to the degree to which a difference between the lifetimes of localized and pseudolocalized vibrations persists.

The description using the Stokes loss function $p(\omega) = \frac{1}{2} a^2(\omega) \cdot \rho(\omega)$ is exact. If we consider that it also includes information about the Stokes losses of localized vibrations then it determines (within the framework of our basic model) the complete shape of the vibronic band and $p(\omega)$ can be found from the experimental band shape (see [75, 76]).

Below we give (without a calculation) the first three moments of the intensity distribution in a peak corresponding to a narrow maximum in the background which arises in the presence of a single narrow vibration band or from a single narrow maximum in $p(\omega)$. The formulas are obtained in the Condon approximation assuming that there is no change in the normal coordinates during the electronic transition but there is a change of the vibration frequencies during the electronic transition.

The zeroth moment $S_0^k(T)$ (integrated intensity of the k-th peak) is

$$S_0^k(T) = \left(\frac{p_1}{p_2} \right)^{k/2} I_k(2\sqrt{p_1 p_0})\exp\left\{ -\sum_\varkappa 2 p_\varkappa \left(\bar{n}_\varkappa + \frac{1}{2} \right) \right\}. \qquad (15.2)$$

The first and second moments $S_1^k(T)$ and $S_2^k(T)$ are calculated assuming $n_\varkappa = n$ for all of the vibrations in the narrow band under consideration, i.e., we neglect differences in the degree of thermal excitation of the oscillators whose frequencies lie within the narrow band. If the band width is narrow in comparison to the average frequency in it this approximation is good.

The first normalized moment is

$$\frac{S_1^k(T)}{S_0^k(T)} = \varepsilon_0 + kh\bar{\omega} + \sum_\varkappa h\Delta\omega_\varkappa \left(\bar{n}_\varkappa + \frac{1}{2}\right). \tag{15.3}$$

The second normalized central moment is

$$\frac{\overline{S_2^k(T)}}{S_0^k(T)} = [\overline{\omega^2} - (\bar{\omega})^2][k + 2p \sqrt{\bar{n}_\varkappa(\bar{n}_\varkappa + 1)} \; \frac{I_{k+1}(2p\sqrt{\bar{n}_\varkappa(\bar{n}_\varkappa + 1)})}{I_k(2p\sqrt{\bar{n}_\varkappa(\bar{n}_\varkappa + 1)})}]. \tag{15.4}$$

Here ε_0 is the energy of the purely electronic transition, $p_\varkappa$ is the Stokes loss in the $\varkappa$-th oscillator in units of the vibrational quantum $h\omega_\varkappa$, $\Delta\omega_\varkappa$ is the change in the frequency $\omega_\varkappa$ during the electronic transition, $\bar{n}_\varkappa$ is the average number of quanta (phonons) in the $\varkappa$-th oscillator at a temperature T,

$$p \equiv \sum_\varkappa{}' p_\varkappa; \; p_0 \equiv \sum_\varkappa{}' p_\varkappa \bar{n}_\varkappa; \; p_1 \equiv \sum_\varkappa{}' p_\varkappa(\bar{n}_\varkappa + 1); \; \bar{\omega}^m \equiv \sum_\varkappa{}' \omega_\varkappa^m p_\varkappa / p;$$

the prime on the summation means that the summation runs only over the narrow band, $\bar{\omega}^m$ is the normalized m-th moment of $p(\omega)$ for the band, and $I_k(x)$ is the k-th order modified Bessel function.

A comparison of these formulas with the results for localized vibrations supports the qualitative conclusions obtained above and permits us to reach some new conclusions.

1. Localized vibrations and δ-function singularities in $p(\omega)$ make identical contributions to the quasiline spectrum: the effect on the characteristics and the temperature dependences of all of the quasilines is the same.

2. Finite-width peaks in $p(\omega)$ can be compared to finite-width vibrational levels of localized oscillators which have finite lifetimes. The finite widths of the peaks cause a finite width (we neglect the radiation width) and additional broadening of all of the quasilines with increasing temperature.

3. Differences in the sizes of the vibration quanta observed in quasiline absorption and emission spectra are not a conclusive argument in favor of the presence of localized vibrations at an impurity center since $p(\omega)$ can have narrow maxima located at different frequencies ω_1 and ω_2, depending on the electronic state. In other words, a change in the frequency of a pseudolocalized vibration during the electronic transition can play the same role as a finite change in a localized vibration frequency.

4. The presence of narrow bands or maxima in the function $p(\omega)$ strongly affects the purely electronic line too (Eqs. (15.2)-(15.4) with k = 0). At medium and high temperatures where an appreciable fraction of the vibrations in a narrow band are excited, transitions in which a phonon is created and another which has nearly the same frequency (differing by less than $\delta\omega$ determined by the possibilities of the experiment) is simultaneously annihilated, begin to play a role. These transitions can lead to a strong increase of the background in the vicinity of the quasilines with increasing temperature. This fact must be taken into account in discussing the experimentally observed temperature dependence.

§ 16.   Characteristic Patterns of

Vibrational Structure

It is not difficult to see that among the numerous theoretically possible vibrational structures corresponding to the basic model, there are four particular cases which allow us to interpret many typical observed spectra.

Case 1.   Replicas of the basic shape well separated, sharp structure due to band vibrations ("ideal case").

We limit ourselves initially to one localized vibration. We have seen in section 13 that creation of λ localized vibration quanta shifts the basic shape $I_\varkappa(\omega, T)$ (i.e., the zero-phonon line and the vibrational background) by a frequency $\lambda\Omega$. The whole intensity in the λ-th replica $(\lambda = 0, 1, 2, \ldots)$ of the basic shape is multiplied by a factor $\exp(-p)p^\lambda/\lambda!$ (p is the dimensionless Stokes loss).

We now consider the case where the frequency Ω of the localized vibration is greater than the width of the basic shape $I_\varkappa(\omega, T)$. Favorable conditions for this case occur with high

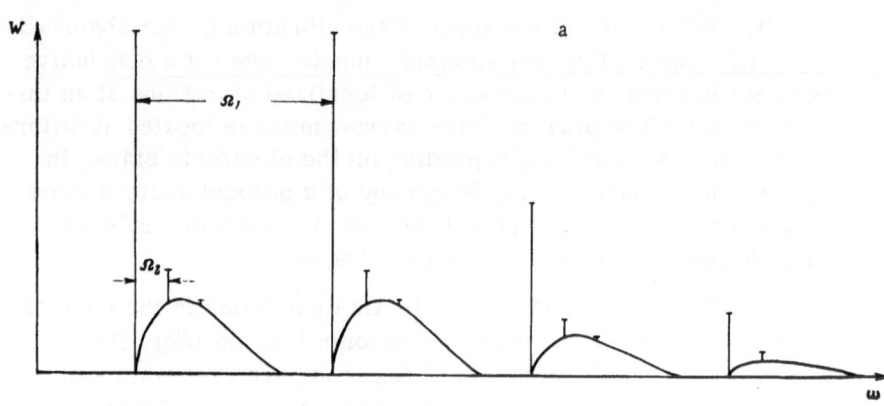

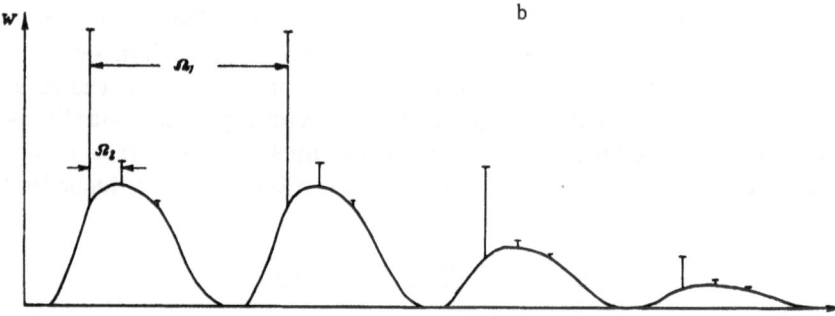

Fig. 16. Theoretical pattern of the quasiline spectrum in the "ideal case." The basic shape and its replicas at intervals of the localized phonon frequency Ω_1 are still distinct but the replicas at intervals Ω_2 overlap (a) absolute zero; b) $T_1 > 0$).

localized vibration frequencies, low Stokes loss in the band vibrations, and low temperatures. In this case the replicas of the basic shape are well separated from each other and the intensity distribution in the spectrum has the form shown in Fig. 15.

If in addition to a high-frequency (Ω_1) localized vibration we have another localized vibration with a lower frequency Ω_2, and its Stokes loss is not large, we obtain the spectrum shown in Fig. 16. It corresponds to strong superposition of replicas of the basic shape $I_\varkappa(\omega, T)$ shifted relative to each other by $\lambda_2\Omega_2$, but the regions of the spectrum shifted by a frequency $\lambda_1\Omega_1$ are still quite sharp and distinct. The pattern obtained can be interpreted as a similar replication at intervals $\lambda_1\Omega_1$ ($\lambda_1 = \pm 0, 1, 2, \ldots$) of a new basic shape $I_{\lambda_2\varkappa}(\omega, T)$ which expresses the vibrational structure

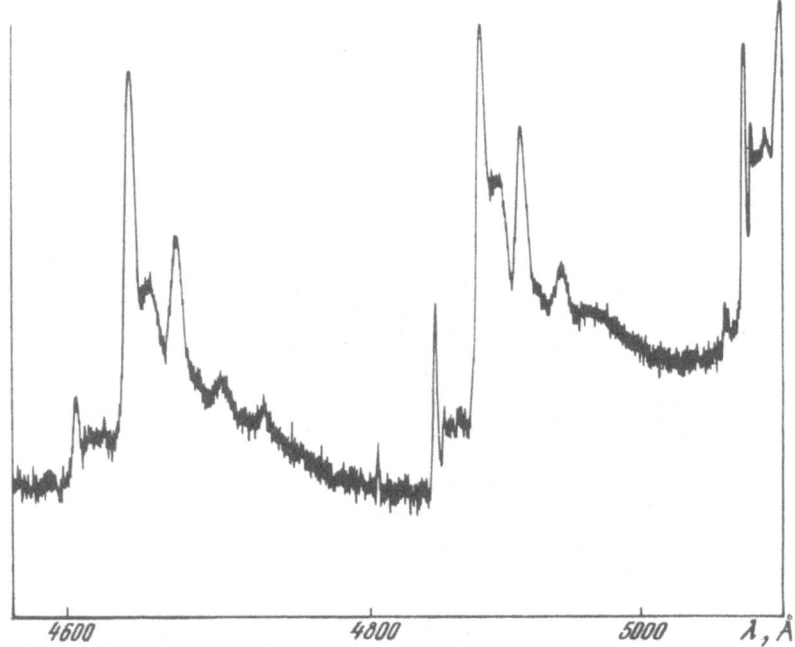

Fig. 17. Microphotometer trace of part of the emission spectrum of $KBr-O_2^-$ at 4.2 K containing two vibrational groups (λ = 4651 and 4888 Å, numbers 5 and 6 in Fig. 18). The similarity law is fairly well fulfilled.

taking the band vibrations and the lower-frequency (Ω_2) localized vibration into account. In the previous section we have shown that such a pattern is obtained if, instead of a localized vibration with a frequency Ω_2, we have a singularity in the vibrational spectrum. The spectra shown in Figs. 15 and 16 are good illustrations of the "similarity law." Spectra of this kind are the "ideal case" of quasiline vibrational spectra.

The low-temperature luminescence spectrum of the O_2^- molecule in an alkali halide crystal shown in Fig. 17 is an example of the "ideal case." The role of the high-frequency localized vibration is played by an intramolecular vibration (modified by the crystalline surroundings) of the O_2^- ion which has a frequency $k_1 = 2\pi\Omega_1/c \approx 1100$ cm^{-1}. It is easily seen that the similarity law is fairly well satisfied. In this case the vibrational structure arises from the interaction of the electronic transition with vibrations significantly distorted by the presence of the O_2^- impurity [81].

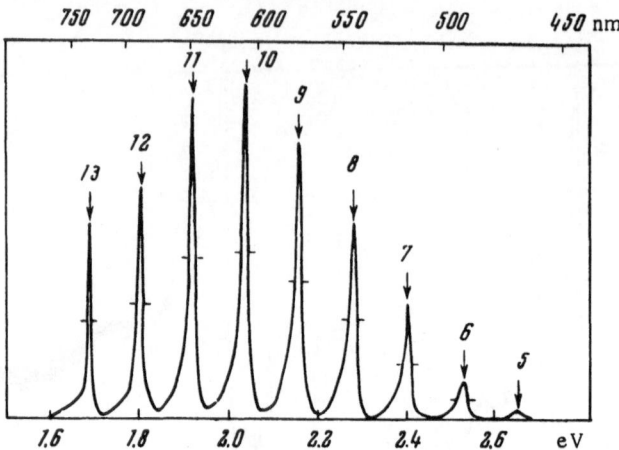

Fig. 18. Luminescence spectrum of $KBr-O_2^-$ at liquid nitrogen temperature. The vibrational structure is not resolved but the bands corresponding to various $0 \rightarrow f$ transitions of the localized vibration are still well separated.

Case 2.   Replicas of the basic shape well separated, detailed structure due to band vibrations not resolved.

An example is the luminescence spectrum of the O_2^- molecule in KBr taken at liquid nitrogen temperature (Fig. 18). It appears in the presence of a high-frequency localized vibration under conditions where the basic vibration spectrum is without detailed structure. If the interaction is with acoustic vibrations which are not strongly distorted by the impurity (no pseudolocalized vibrations in the acoustic branch) there is no structure in the background at helium temperatures (or even at absolute zero).*   Another more common possibility is that with increasing temperature the vibrations are already strongly excited and the zero-phonon line and singularities in the background lose intensity and broaden so that the structure vanishes. At this temperature the localized vibration is still not excited or only slightly excited and the structure associated with it persists.†   Just this possibility is realized in $KBr-O_2^-$.

---

* A narrow optical vibration branch can play the role of a localized vibration here, giving a series of replicas of the basic spectrum. Such a situation occurs for "bound excitons" in CdS [30].

† A criterion for the degree of excitation of the localized vibration could be the half-

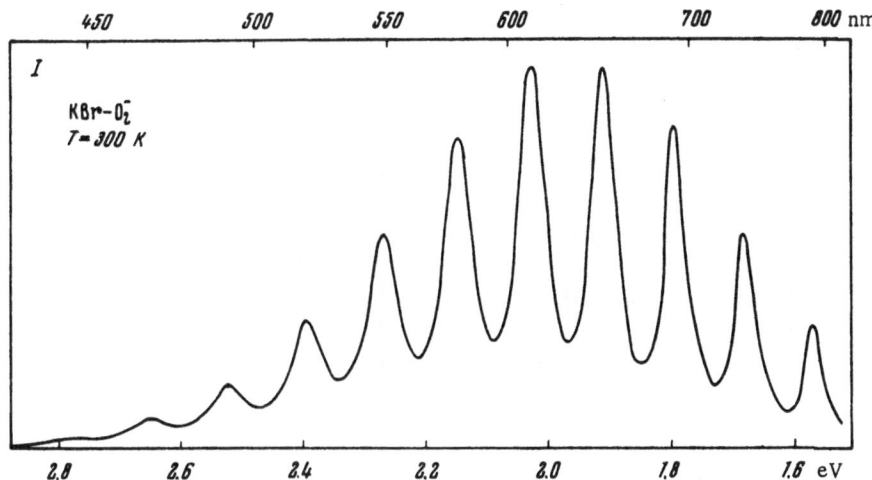

Fig. 19. Luminescence spectrum of $KBr-O_2^-$ at room temperature. There is already overlapping of the bands corresponding to various localized vibrational transitions.

It is also evident that the appearance of detailed vibrational structure depends on the capabilities of the apparatus and the conditions of the experiment. For example, the O_2^- spectrum in alkali halide hosts at 4.2 K, which we cited above as an example of the "ideal case," was measured with less resolution in [82] and therefore it was observed with much less detail.

The structure of the spectrum in case 2 contains substantially less information about vibrations of the crystal and impurity and their interaction with the electronic transition than the "ideal case." It should be stressed, however, that due to the clear resolution of the individual localized vibration bands, spectra of this kind can be the basis for fairly well-established and quite detailed theoretical analyses. Thus the areas of the individual bands (zeroth moments) give valuable information about the Franck–Condon factors for the localized vibration, from which we can construct the potential curve of the localized vibration in the excited

---

width (in the presence of structure, the envelope shape taking account of the weights of the peaks and the distances between them) of the whole vibronic band for a given electronic transition. The spectrum broadens according to the degree of thermal excitation of the localized vibration in the electronic ground state, and in the luminescence spectrum according to the degree of excitation in the electronic excited state.

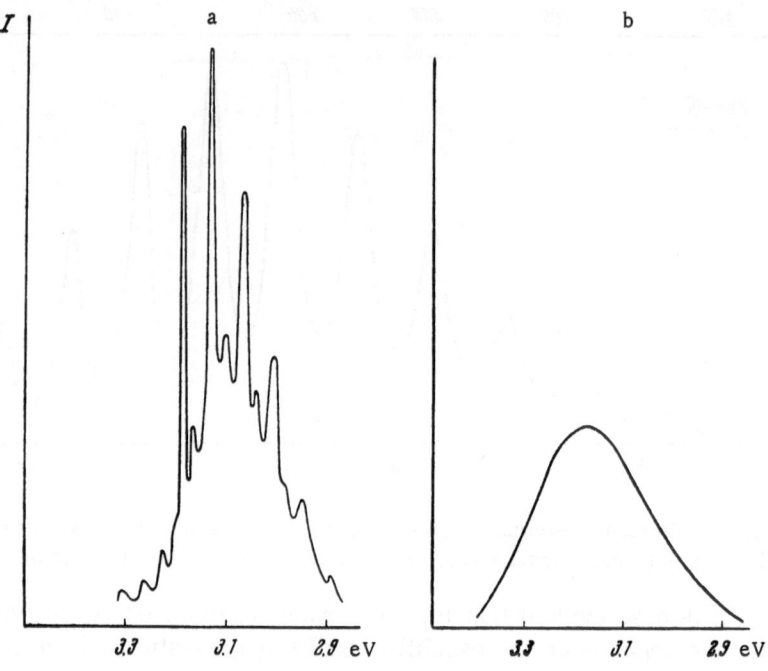

Fig. 20. Emission spectrum of CaO−Bi measured at 100 K (a) and at 295 K (b) [87].

electronic state by a finite calculation [83]. To find the average characteristics of the interaction with the band vibrations we can use the higher-order moments and their temperature dependences determined from the experimental shapes of the individual bands, since the method of moments is fully applicable to them too if we disregard the interaction between the localized vibrations and the band vibrations [84-86, 8].

Case 3. Peaks of the basic shape still well resolved but appreciable overlap of the wings; detailed band vibrational structure not resolved ("glove-shaped" spectrum).

The luminescence spectrum of O_2^- in KBr measured at room temperature (Fig. 19) can serve as an example. The general structureless background arises from mutual overlap of the vibrational wings of the various images of the basic shape. Another characteristic example is the luminescence spectrum of Bi^{3+} in CaO (Fig. 20a) [87]. While the series of bands arises from local-

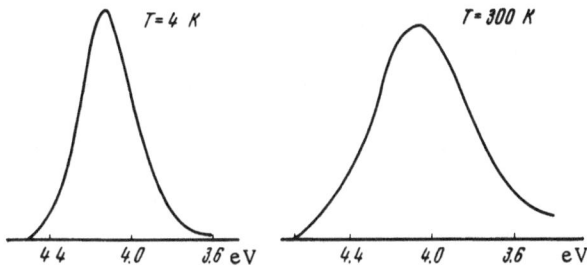

Fig. 21. Ultraviolet emission band in KCl−Tl at 4 and 300 K
[91].

ized vibrations in the KBr−O_2^- system, in CaO−Bi in all likelihood
it arises from pseudolocalized vibrations. The spectrum of case 2
considered above turns into a "glove" spectrum due to the increase
in the multiphonon transition probability with temperature and the
related broadening of the vibrational wings. Evidently systems
which have case 3 spectra are often encountered in which case 2
spectra do not appear with decreasing temperature. This is understandable, since to get fully separated images of the basic shape
it is necessary that two rigid conditions be fulfilled: one needs a
quite weak interaction with the band vibrations and a high localized
vibration frequency in comparison to the band vibration frequencies. Due to the impossibility of separating the individual images
of the basic shape, this kind of spectrum gives still less information about the localized crystal dynamics than case 2 spectra.

C a s e 4.   Continuous bell-shaped spectrum. The characteristic form of the spectrum is shown in Fig. 21, and the bell-
shaped absorption and luminescence bands of KCl activated by thal-
lium ions can serve as an example. In systems of the KCl−Tl
type, in which we can assume a strong interaction with a quite
broad range of band vibrations, vibrational structure is not observed even at helium temperatures. On the other hand, a broad
band close to gaussian in shape is the high-temperature limit of
the vibronic spectrum for any system. However, this limit is
practically unattainable for many spectra due to thermal quench-
ing, thermal disruption of the impurity center, melting of the crystal, etc.

The information in these spectra is already in a highly averaged form. Well-understood methods are available for analyzing

them [10, 8, 88, 89]. Because these lines are nearly gaussian in shape the method of moments, previously used for the whole vibronic band [8, 84–86], is applicable (see Appendix VI).

Exercise 21.   Can we assume that the widths of the individual bands in case 2 are due to the finite lifetimes of the corresponding levels of the localized oscillator or pseudolocalized vibration?

Exercise 22.   Assuming that the half-width $\Delta\Omega$ of an individual band in a case 2 spectrum is due to the lifetime τ of a pseudolocalized vibration, find the relation between τ/T and $\Delta\Omega/\overline{\Omega}$, where T and $\overline{\Omega}$ are the period and the frequency of the pseudolocalized vibration.

Concluding Remarks

As we have already stressed, in general vibronic structure is quite varied even for the basic model.

The four limiting cases which we have discussed are only seldom realized in pure form. In real spectra the vibronic bands corresponding to different impurity centers or other electronic transitions in the same center quite often overlap.

In a single individual vibronic band it often happens that vibrational structure appears in one part but not in others. Evidently the vibrational structure is most distinctly manifested in the neighborhood of the purely electronic line. In [13] a formula was given which describes the vibrational structure in the background of a complicated band for a substantial Stokes loss. The structure is represented as a "modulation" of the smooth bell-shaped curve by oscillatory factors whose amplitudes decrease with distance from the purely electronic line. It is fairly reliable to seek vibrational structure in the region of the purely electronic line even when the Stokes loss is substantial, even in bands measured at helium temperatures where no vibrational structure has yet been observed. The structure may be observed in a more detailed investigation.

As an example of the importance of knowing just where in the spectrum to look, we mention the observation of vibrational structure in the vicinity of the purely electronic line in such traditionally structureless spectra as the luminescence band of the R center in LiF and NaF crystals [43, 44].

From our analysis of the theory developed for the basic model of the interaction with vibrations (and the comparison with experiment) it can be seen that even in this version the theory suffices to explain the basic experimental data. We will return to the corrections introduced by better theories, which can sometimes be quite important, in chapter IV.

§ 17.   Summary of the Results for the Basic Model.   The Need for Improved Theories

We will briefly reiterate the results of the theoretical analysis for the basic model.

As the basic model for describing lattice vibrations and their interaction with electronic transitions we have used a model having the following properties: (1) the vibrations of the crystal containing impurities are harmonic, (2) we consider dispersion of the band vibrations; some of the normal modes can be localized, (3) the oscillator strength (absolute square of the matrix element) of the electronic transition does not change when the lattice vibrates, and (4) the change of the electronic state of the impurity leads only to a change in the equilibrium positions of the normal modes, i.e., their frequencies do not change, nor is there a rotation of the normal coordinate system.

Despite the dispersion of the band vibrations and even with a strong interaction between the electronic transition and the band vibrations, our model has a resonant purely electronic line of zero width, described by a δ-function, in the absorption and luminescence spectra. Neither the magnitude of the coupling with vibrations, characterized by expressing the Stokes loss in terms of the number of vibrational quanta, nor the temperature affects the width of the line or its integrated intensity $I(T)$. With increasing Stokes loss or with increasing temperature the integrated intensity of the purely electronic line rapidly decreases. For example, with a Stokes loss of n effective vibrational quanta the purely electronic line has only e^{-n} times the whole intensity of the absorption or luminescence band which corresponds to a given electronic transition at absolute zero. At high temperatures $I(T)$ also decreases exponentially with T. Since the local lattice dynamics about the

impurity center affect the vibrational properties of the spectrum
[40], the temperature dependence is characteristic not only of the
host crystal but also sharply reflects the properties of the impurity center.

The purely electronic line is a quasiline, whose components
correspond to transitions occurring from various initial vibrational states with $t > 0$. These transitions are zero-phonon transitions
in which the vibrational state of the impurity and its surrounding
lattice do not change (phonons are neither created nor annihilated).

When there is a localized vibration of frequency Ω the spectrum contains vibrational replicas of the purely electronic line at
intervals $k\Omega$ (k is an integer). In this approximation to the theory
they too are represented by δ-function peaks. The distribution of
integrated intensity over the series of peaks at temperatures
where the localized vibrations are still not excited is given by
$(p_\lambda^k/k!)I(T)$, where p_λ is the dimensionless Stokes loss for the λ-th
localized vibration, and $I(T)$ is the intensity of the purely electron-
ic line.

Transitions accompanied by changes of the states of band
vibrations lead to formation of a vibrational background in the
spectrum (the analog of the wings in the Mössbauer spectrum, see
chapter III). This background can have quite distinct structure
and even peaks due to interactions with band vibrations of the limiting frequencies or with pseudolocalized vibrations. The peaks in
the background already have finite widths and the widths are determined by the composition of the band vibration wave packet occurring in the process. The integrated intensities in the series of
peaks which are images of the purely electronic line at intervals
$k\Pi$ (k is an integer and Π is the pseudolocalized vibration frequency) are also described approximately by the formulas above.

The vibrational wings, together with the purely electronic
line, are replicated according to a "similarity law" at intervals
$k\Omega$ (or $k\Pi$) where the intensity distribution is given by factors
$S_{0\varkappa}^k(T)$ (the Franck–Condon factor) according to (13.11) and (15.2).

It is clear that in reality the purely electronic line and its
vibrational replicas have finite widths. But since the approximations made in the basic model are fairly well based, these lines,
in particular the purely electronic line, in fact should be quite

narrow. The width of the line is at least the radiation width due to the finite lifetime of the excited electronic state for spontaneous emission. To find it we trivially complete the calculation of the interaction of the impurity with the radiation field (using, for example, the Wigner-Weisskopf formula which takes radiation damping into account [90]).

The radiation width for an allowed electronic transition with a lifetime of $\tau \approx 10^{-8}$ sec amounts to about 10^{-3} cm^{-1} and is less for a forbidden transition. This width is actually very small in comparison to the electronic transition frequencies. Even very small relative broadening (i.e., broadening in relation to the average vibrational quantum) can give broadening three or four orders of magnitude greater than the radiation width. Therefore the quasiline is very sensitive to many fine details of the vibronic interaction not taken into account in the basic model. The role of these interactions, particularly their effects on the linewidth, is important.

One must add anharmonic interactions between the vibrations to the basic model, and consider changes of the normal coordinate system (rotation of the normal coordinate axes) during the electronic transition. We will see that the first factor leads to a substantial broadening of the vibrational replicas of the purely electronic line and also to a radical change in the integrated intensity of the purely electronic line in the presence of localized vibrations. The clearest manifestation of the second factor is the thermal broadening of the purely electronic line.

Furthermore, it is necessary to investigate the role of the dependence of the oscillator strength of the electronic transition on the vibrational motion. This effect turns out to be important for transitions forbidden by spatial symmetry in the equilibrium configuration of the impurity center.

In conclusion we reiterate that real crystals are not always ideally homogeneous. We can estimate the effect of the inevitable inhomogeneity of the host crystal and the isotopic composition of the impurity and its surroundings. It turns out that these factors can quite sharply affect the observed properties of the quasilines.

In discussing the role of these improvements it is useful to compare their effects with the analogous factors in the Mössbauer

effect.   Therefore we now turn to the Mössbauer effect and to the
analogy between this effect and vibronic transitions.

Chapter 3

The Relation between Vibronic Spectra and Mössbauer Spectra

§ 18. The Mössbauer Effect

We will give a brief survey of the basic physics of the Mössbauer effect. Attention will be focused on aspects of interest for a comparison with vibronic spectra. More detailed information about this interesting and important effect can be found in the small books of Frauenfelder [92] and Gol'danskii [93] intended for a wide readership of physicists and chemists.

The name Mössbauer effect is given to elastic emission and absorption of gamma quanta by atomic nuclei bound in solids; it is elastic in the sense that it is not accompanied by any change of the internal energy of the body, i.e., by emission or absorption of phonons [94]. This effect was discovered by the German physicist Mössbauer in 1958. The elastic transitions (transitions without a change in the vibrational state of the crystal lattice) correspond to very narrow gamma-ray emission and absorption lines which have the natural width $\Gamma = h/\tau$ and a frequency $\omega_0 = E_{00}/h$ where τ is the average lifetime of the excited nuclear state which participates in the gamma transition and the energy E_{00} is the energy difference between the initial and final nuclear states.

The small width of the Mössbauer line, $h\Delta\omega \approx 10^{-5}$ to 10^{-10} eV, and the unprecedentedly small relative width $\Delta\omega/\omega \approx 10^{-10}$ to 10^{-15}, allows its use for measuring exceedingly small relative

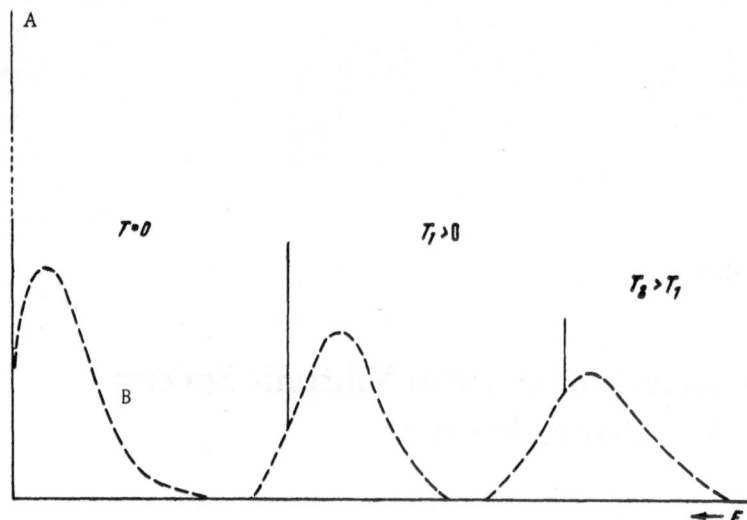

Fig. 22. Theoretical energy distribution in the spectrum of gamma quanta emitted from a nucleus in a crystal. A) Mössbauer line (zero-phonon line); B) vibrational wing.

shifts in the gamma quantum frequency due to various interactions of the emitting or absorbing nucleus, or of the gamma quantum.

In addition to the elastic transitions there are evidently inelastic transitions, too, i.e., processes in which the change in the internal state of the nucleus and emission (absorption) of the gamma quantum is accompanied by a change in the vibrational state of the crystal. In other words, there are transitions accompanied by creation or annihilation of a certain number of phonons (and localized vibration quanta) in the crystal. These transitions correspond to lines displaced from ω_0 by a distance determined by the (algebraic) sum of the energies of the phonons which participate in the transition.

The spectrum of gamma quanta absorbed or emitted by an atomic nucleus in a crystal lattice has the general features shown in Fig. 22. With increasing temperature the probability of transitions in which phonons participate increases. Note that the wings shown in Fig. 22 in the Mössbauer spectrum are only known theoretically. It has not yet been possible to measure the wings in the spectrum in experimental investigations of the intensity and shape of the Mössbauer line.

From what has been said, and especially from a comparison of Fig. 22 with Figs. 11 and 12, one infers that there are certain analogies between Mössbauer spectra and vibronic spectra. The Mössbauer line corresponds to a purely electronic line and the vibrational wings correspond to the vibrational background around the purely electronic line.

The purely electronic line and the Mössbauer line (which could also be called a purely nuclear line) are both zero-phonon lines. In a word, the analogy is clear and not surprising, so that after the Mössbauer effect became well known to a wide circle of physicists these analogies were established or mentioned independently and more-or-less simultaneously in several papers.

The first work in which the analogy between the purely electronic line and the Mössbauer line was explicitly formulated was that of Trifonov [26].* This analogy was also mentioned in a less explicit form in the work of Hopfield [42] and Silsbee [95].

In [27, 72] the analogy was established taking account of localized vibrations, on the basis of a comparison of the Mössbauer effect with the Shpol'skii effect [22]. In [38] the analogy was considered in more precise versions of the theory and in [96] the general case was worked out.

In order to analyze the reasons for the analogy between the spectra we turn to the mechanism of the interaction of the gamma quanta emitted (absorbed) by the nucleus with lattice vibrations and compare it to the interaction mechanism in vibronic transitions.

At first glance the vibrational interaction processes in the Mössbauer effect and in vibronic spectra are essentially different and one might think that the vibrational structure in the gamma ray spectrum and visible light emitted in these processes is essentially different. Actually, in the Mössbauer effect the interaction with vibrations arises because the radioactive nucleus itself takes up the recoil momentum of the gamma quantum, thus receiving a certain excess recoil kinetic energy. A more detailed treatment, of which we give the simplest version in the next section, shows that the vibrational structure of the spectrum is determined

---

* Trifonov set forth his work in a seminar of the theoretical physics division of the Physics Institute at Leningrad State University in March, 1961.

by the way this energy is distributed among the normal modes of the crystal.

In a vibronic transition the impurity atom emits a visible light quantum, whose recoil momentum is only a thousandth of the recoil momentum of a gamma quantum. Therefore the recoil energy can be neglected in this process. But as a result of the transition of the electron to another orbit there is a substantial change in the electric charge distribution in the atom and also the electrical forces which bind the impurity in the crystal, In sum, after a vibronic transition the impurity and its immediate surroundings have an excess potential energy. The detailed treatment carried out in the previous sections of this book shows that the vibrational structure of the spectrum is determined by how this potential energy, namely, the Stokes loss energy, is distributed over the normal modes of the crystal.

Thus in the Mössbauer effect the interaction with vibrations occurs through transfer of kinetic energy to the crystal, while for vibronic transitions potential energy is transferred. It would appear that the basic properties of the resulting spectra might be radically different.

Nevertheless the spectra turn out to be identical in their basic features. The formulas which describe the distribution of energy in the spectra are startlingly similar. There is only one difference: at places where we have the Stokes loss energy and its distribution over normal modes in formulas for vibronic spectra, in formulas for the Mössbauer effect we have the recoil energy and its distribution over the normal modes.

The chief reason for the analogy is that the lattice vibrations are well described by small vibration theory, i.e., in the final analysis the vibrations can be regarded as a set of independent elementary vibrational systems — harmonic oscillators. The harmonic oscillator energy has a remarkable property in that it contains the coordinates and momenta in a symmetrical way. The potential energy of a harmonic oscillator depends quadratically on the coordinates, i.e., in precisely the same way as the kinetic energy depends on the momentum. Because of this particular property the recoil energy and the Stokes loss energy turn out in the final analysis to have the same effect on the structure of the spectrum.

The analogy can be followed also in the improved versions of the theory where the lattice vibrations and their interactions are described on very general assumptions (see section 22).

Further development of the theory has shown that the analogy is useful for rapid and parallel calculations of a number of important properties of the purely electronic line and the Mössbauer line (see [38-40]). Thus, for example, more precise versions of the theory have been worked out which explain the thermal shift and the broadening of the purely electronic line and the Mössbauer line taking anharmonicity of the lattice vibrations into account [37-40], along with scrambling of the normal coordinates, etc. It should be stressed that, due to the extreme narrowness and the resulting high sensitivity of the Mössbauer line, correction factors are just as necessary in the theory of the Mössbauer effect as in the theory of the purely electronic line. The Franck—Condon principle has also been formulated for the Mössbauer effect [97] (see Appendix I).

§ 19. Derivation of the Basic Equations for the Vibrational Structure of the Gamma Transition Spectrum

We now calculate the integrated intensity and temperature dependence of the Mössbauer line in first-order perturbation theory. Strictly speaking, in this problem the first order does not suffice since the width of the observed line is in many cases only somewhat more than the radiation width. In the first order of ordinary perturbation theory we must consider the shape of the spectral line to be a δ-function; to obtain the correct (natural) lorentzian shape one must use second-order perturbation theory (see, for example, [90]).

It turns out that to investigate the shape of the Mössbauer line, as we would expect, it is necessary to go to second-order perturbation theory; however, to obtain the basic formula for the integrated intensity and the vibrational wings the first order is quite sufficient.

Thus we will proceed from first-order perturbation theory (see section 4, part (c)) in which the transition probability W_{if}

from state i to state f is proportional to the absolute square of the matrix element of the perturbation operator $\hat{P}$ which gives rise to the transition

$$W_{if} \sim |P_{fi}|^2, \quad P_{fi} = \int \Psi_f^*(\tau)\, \hat{P} \Psi_i(\tau)\, d\tau, \tag{19.1}$$

where τ denotes the set of all coordinates which describe the system.

We will first establish the wave function and the perturbation operator for the Mössbauer effect.

When a gamma quantum is emitted by a radioactive nucleus in a crystal lattice the following things change: the number of photons in the electromagnetic field (a gamma quantum is created), the internal state of the nucleus, the vibrational state of the lattice, and also the state of the electronic shells (due to the change in the nuclear state). The wave function of the system should also describe all of these states and should thus depend on the nuclear coordinates ρ (ρ is the set of nucleon coordinates), the electronic coordinates r, and the lattice vibrational coordinates R. It is now natural to turn to the adiabatic approximation. However, it is evident that we now want to use it twice, since the internal motion in the nucleus is quite rapid in comparison to the electronic motion, which in turn is rapid in comparison to the vibrational motion of the lattice. In this approximation the wave function for the whole system takes the form

$$\Psi_{kli}(\rho;\ r,\ R) = \chi_k(\rho;\ r,\ R)\, \Phi_{kl}(r,\ R)\, \varphi_{kli}(R), \tag{19.2}$$

where k, l, and i are the quantum numbers of the internal nuclear, electronic, and vibrational states respectively. The dependence of the nuclear wave function χ_k on the electronic and vibrational motion and the dependence of the electronic wave function on the vibrational coordinates are parametric (see section 1).

We use the usual interaction between a plane electromagnetic wave and a system of charges for the interaction operator between the gamma quantum and the system (see, for example, [70, 90], and Appendix V)

$$\hat{P} = \sum_n \frac{Q_n}{m_n}(\varepsilon p_n) \exp(-i\, K r_n), \tag{19.3a}$$

where the index n denotes all of the charged particles in the system, p_n and r_n are the momentum and coordinate operators of the n-th particle, Q_n and m_n are the charge and mass of the n-th particle, and K and ε are the wave vector and polarization vector of the photon (gamma quantum). In our case, where the charge can be divided into the electric charge of the nucleons, electrons, and atomic nuclei (or ions) the operator $\hat{P}$ can be written as

$$\hat{P} = \sum_{\lambda} \frac{Q_{\lambda}}{m_{\lambda}} (\varepsilon \pi_{\lambda}) \exp(-i K \rho_{\lambda}) + \sum_{l} \frac{Q_{l}}{m_{l}} (\varepsilon p_{l}) \exp(-i K r_{l})$$

$$+ \sum_{a} \frac{Q_{a}}{m_{a}} (\varepsilon P_{a}) \exp(-i K R_{a}), \tag{19.3b}$$

where the index λ enumerates the nucleons, l denotes the electrons ($Q_l = -e$, where e is the unit charge) and α denotes the nuclei of the atoms or the ions in the crystal lattice.

The gamma transitions of interest to us* are given by the first term in (19.3b). The second term of the perturbation operator leads to vibronic transitions, and the third, to vibrational transitions.

Thus in our case we substitute the wave function (19.2) and the first term of the perturbation operator (19.3b) into the matrix element (19.1). Finding the probability of a transition $kli \rightarrow k'l'i'$ reduces to calculating the integral

$$P_{k'l'i';\,kli} = \int \chi_{k'}^{*} (\rho;\ r,\ R)\, \Phi_{k'l'}^{*} (r,\ R)\, \varphi_{k'l'i'}^{*} (R)$$

$$\times \left\{ \sum_{\lambda} \frac{Q_{\lambda}}{m_{\lambda}} (\varepsilon \pi_{\lambda}) \exp(-i K \rho_{\lambda}) \right\} \chi_{k} (\rho;\ r,\ R)\, \Phi_{kl} (r,\ R)$$

$$\times \varphi_{kli} (R)\, d\rho\, dr\, dR. \tag{19.4}$$

Here we integrate over the whole range of the coordinates ρ, r, and R. In the final analysis we are only interested in the vibrational structure of the spectrum; therefore we should free ourselves from extraneous complications in calculating (19.4) by introducing empirical parameters analogous to the electronic matrix element in the theory of the vibrational structure of electronic

---

* More precisely, we speak of "nuclear–electronic–vibrational transitions" since the change in the nuclear state is accompanied by a change in the vibrational and (strictly speaking) the electronic state.

transitions. We proceed to do this as in section 4, and begin the integration with the coordinates of the fastest subsystem, i.e., with the internal nuclear (nucleon) coordinates. We have

$$P_{k'l'i'; \, kli} = \iint dR \, dr \int d\rho \chi_{k'}^{*}(\rho; \, r, \, R) \Big\{ \sum_{\lambda} \frac{Q_{\lambda}}{m_{\lambda}}$$

$$\times (\varepsilon \pi_{\lambda}) \exp(-i \, \boldsymbol{K} \rho_{\lambda}) \Big\} \chi_{k}(\rho; \, r, \, R) \dot{\Phi}_{k'l'}(r, \, R) \dot{\varphi}_{k'l'i'}(R)$$

$$\times \Phi_{kl}(r, \, R) \varphi_{kli}(R) \equiv \iint dR \, dr \, d_{k'k}(r, \, R) \dot{\Phi}_{k'l'}(r, \, R)$$

$$\times \Phi_{kl}(r, \, R) \dot{\varphi}_{k'l'i'}(R) \varphi_{kli}(R). \tag{19.5}$$

Here we simply introduce a notation for the integral over the internal nuclear coordinates.

$$\int \chi_{k'}^{*}(\rho; \, r, \, R) \Big\{ \sum_{\lambda} \frac{Q_{\lambda}}{m_{\lambda}} (\varepsilon \pi_{\lambda}) \exp(-i \, \boldsymbol{K} \rho_{\lambda}) \Big\} \chi_{k}(\rho; \, r, \, R) \, d\rho$$

$$\equiv d_{k'k}(r, \, R).^{*} \tag{19.6}$$

Despite the fact that the internal nuclear motion only senses the motion of the electrons on the atoms and the vibrational motion of the lattice to a vanishingly small degree, we cannot say that the matrix element $d_{k'k}(r, R)$ does not depend on r and R. Because the coordinates ρ, r, and R have been chosen in the laboratory coordinate system, they also inherently contain the motion of the nucleus as a whole in this coordinate system. In other words, the wave function $\chi(\rho; r, R)$, in addition to the internal nuclear motion, describes the participation of the nucleus as a whole in the vibrational motion of the crystal lattice. In order to avoid this undesirable complication we write the internal nuclear coordinates of the gamma-active nucleus in the center of mass system of the nucleus, i.e., we introduce the new variables

$$\rho' = \rho - \boldsymbol{R}_{\alpha}, \tag{19.7}$$

into (19.6), where $\boldsymbol{R}_{\alpha}$ is the position vector of the gamma-active nucleus in the laboratory coordinate system (ρ and ρ' are multi-dimensional vectors). We can rewrite (19.6) in the form

---

* Evidently in introducing the parameter $d_{k'k}(r, R)$ it is not necessary to choose the operator for the interaction between the gamma quantum and the internal nuclear degrees of freedom in the form (19.3ab). It is sufficient that it depend only on the internal nuclear coordinates ρ and this requirement is automatically fulfilled if the perturbation operator, as is usually the case, is a bilinear combination of the coordinates of the electromagnetic field and matter. Thus it is evident that we should think of the collective coordinates for nuclear motion in ρ

$$d_{k'k}(r,\ R) = \int \chi_{k'}^{*}(\rho';\ r,\ R)\left\{\sum_{\lambda}\frac{Q_{\lambda}}{m_{\lambda}}\ \exp\left[-i\,K\,(\rho_{\lambda}' + R_{a})]\,(\varepsilon\pi_{\lambda})\right\}\chi_{k}(\rho';\ r,\ R)\,d\rho',$$

where the new χ_k describes the internal nuclear state in the nuclear center-of-mass system. If we remove the part which does not depend on ρ' from under the integral sign, we find

$$d_{k'k}(r,\ R) = e^{-iKR_a}\int \chi_{k'}^{*}(\rho';\ r,\ R)\left\{\sum_{\lambda}\frac{Q_{\lambda}}{m_{\lambda}}\exp\left(-iK\rho_{\lambda}'\right)\right.$$

$$\left. \times(\varepsilon\pi_{\lambda})\right\}\chi_{k}(\rho';\ r,\ R)\,d\rho' \equiv e^{-iKR_a}D_{k'k}(r,\ R), \tag{19.8}$$

where the matrix elements $D_{k'k}$ already are very weakly dependent on the electronic and vibrational coordinates.*

The quantity $|D_{k'k}(r,\ R)|^2$ gives the probability of a gamma transition in the atomic nucleus. By analogy with the corresponding electronic quantities it will be called the oscillator strength of the internal nuclear transition or the "internal nuclear" matrix element. Furthermore, the matrix element (19.6) can be expanded in a Taylor series and one can state the theory not in terms of the function $D_{k'k}(r,\ R)$ but of the constant coefficients in the series. Since nucleon motion inside the nucleus is little sensitive to the states of the electronic shells of the atoms or to the lattice vibrations, the approximation

$$D_{k'k}(r,\ R) \approx D_{k'k} = \text{const} \tag{19.9}$$

has a much better basis than the fairly well established Condon approximation in the theory of vibronic transitions.† We thus have for the matrix element of the "internal nuclear−electronic−vibrational transition"

---

* The appearance of the factor e^{-ikR_a} is related to the choice of the perturbation in the form (19.3a), but a more careful treatment shows that only the phase factor in the operator (19.3a) is important. In the final analysis the phase factor appears because the wave function of the photon is a plane wave.

† It does not follow that it makes no sense to consider deviations from the approximation (19.9) in the theory of the Mössbauer effect. In particular, we cannot consider *a priori* that the effect of the electronic and vibrational motion cannot make a perceptible contribution for forbidden gamma transitions. There is also "slow" internal motion, whose coupling with the electrons and vibrations sometimes must be considered (spin degeneracy, splitting of quadrupole sublevels, etc.) Some estimates have been given in [98].

$$P_{k'l'i';\,kli} = D_{k'k} \iint dR\,dr\,\Phi^*_{k'l'}(r,\ R)\,\Phi_{kl}(r,\ R)\,\varphi^*_{k'l'i'}(R)\,\varphi_{kli}(R)\exp(-i\,\boldsymbol{K}\boldsymbol{R}_a),$$

$$(19.10)$$

where $\boldsymbol{R}_a$ is the coordinate of the center of mass of the gamma active nucleus, which belongs to the set of coordinates $\boldsymbol{R}$. The electronic wave functions Φ_{kl} and $\Phi_{k'l'}$ belong, strictly speaking, to different sets of orthonormalized functions k and k' since in a gamma transition $k \rightarrow k'$ the hamiltonian of the electrons changes somewhat. However, this change is exceedingly small, like the dependence of the matrix element of the internal nuclear transition on the electronic coordinates, which we neglected. Therefore, to a very good approximation we can say that both of the electronic wave functions belong to the same orthonormal set. Since we have considered gamma transitions in which the electronic state does not change, we set $l = l'$. The integration over electronic coordinates then reduces to the normalization integral; we assume that the electronic functions are normalized and equate the results to unity.

The vibrational function in (19.10) can also be regarded as independent of the indices k and k' of the internal nuclear state to a very good approximation. The electronic state does not change in these transitions $(l = l')$.* In sum we can consider that both vibrational functions belong to the same set of orthonormal functions and characterize them by the same index i (which denotes the set of all vibrational quantum numbers of the lattice).

Thus the distribution of the probability over vibrational levels $i \rightarrow i'$ and gamma transitions $k \rightarrow k'$ occurring in an atomic nucleus in a crystal is given by the absolute square of the matrix element

$$P_{k'l'i';\,kli} = D_{k'k} \int \varphi_{i'}(R)\,e^{-i\boldsymbol{K}\boldsymbol{R}_a}\varphi_i(R)\,dR. \qquad (19.11)$$

To obtain the vibrational structure of the spectrum ("internal nuclear–vibrational" bands) as usual we sum over final and average

---

* In principle there are also "internal nuclear–electronic" transitions in which the gamma transition is accompanied by creation and annihilation of a certain number of quanta of electronic excitation. Due to the extremely weak coupling between the internal nuclear and electronic motions (the "Stokes loss" in the excitation of electronic motion during a gamma transition is very small) the probabilities are vanishingly small.

over initial vibrational states. The Mössbauer line is, just as the purely electronic line, a zero-phonon quasiline. It corresponds to transitions ki → k'i.

Exercise 23. Derive the basic equations for the "double" adiabatic approximation used to describe gamma transitions in the nuclei of atoms embedded in a crystal lattice.

Exercise 24. A gamma transition occurs between energy levels E_1 and E_2, whose finite widths are neglected, in the nucleus of a stationary free atom of mass M.

1. What is the frequency of the emitted or absorbed gamma quantum corresponding to this internal nuclear transition? (Proceed from energy and momentum conservation.)

2. What is the frequency of the emitted or absorbed gamma quantum if the atom moves with velocity v at an angle α to the direction of observation?

3. Calculate the absorption and emission spectra of gamma quanta in an ensemble of free atoms undergoing thermal motion at temperature T.

§ 20. Integrated Intensity of the Vibrational Band in a Gamma Transition

As a preparatory step in studying the Mössbauer line we calculate the integrated intensity (zeroth moment) $S_0(T)$ of the vibrational band in a gamma transition, which is given by Eq. (9.1).

The derivation of Eq. (9.2), based on the general ideas of quantum mechanics, remains valid in this case too. Naturally, to obtain the answer of interest to us we put the explicit form of the perturbation, $D(R) = AD_{k'k}\exp(-iKR_a)$, in (9.2). We have

$$S_0(T) = A \sum_i n_i \int dR\, D_{k'k}^* e^{iKR_a} D_{k'k} e^{-iKR_a} |\, \varphi_i(R)\,|^2 = A \sum_i n_i\, |\, D_{k'k}\,|^2 = NA\,|D_{k'k}|^2,$$

$$(20.1)$$

where $|D_{k'k}|^2$ is the oscillator strength of the gamma transition, A is a constant proportionality factor (see Appendix V), and N is the number of gamma-active nuclei.

Thus the integrated intensity of the "internal nuclear–vibrational band" does not depend on the temperature. This result is analogous to that obtained in section 9 for a vibronic band in the Condon approximation.

We can also calculate higher moments of the vibrational band which arises in the gamma transition. They turn out to be useful for discussing and illustrating the physical ideas of the Mössbauer effect [96, 99].

§ 21. The Integrated Intensity of the

Mössbauer Line and Its Temperature

Dependence

To find the integrated intensity $I(T)$ of the Mössbauer line we must calculate the absolute square of the matrix element $P_{k'i,\ ki}$ according to (19.11) with subsequent summing over final and averaging over initial vibrational states. Rewriting (19.11) setting $i' = i$ and choosing the z axis of the cartesian coordinates system along the wave vector K of the gamma quantum, we find

$$P_{k'i,\ ki} = D_{k'k} \int \varphi_i^2 (R) e^{-iK_z R_{\alpha z}} dR. \tag{21.1}$$

Here R denotes the set of all vibrational coordinates of the lattice, i the set of all vibrational quantum numbers, and R_α the position vector of the gamma-active nucleus. We will assume that the vibrational wave functions are real.

We must calculate the quantity (compare with section 7)

$$I(T) = A \sum_i v_i |P_{ii}|^2 = A|D|^2 \sum_i v_i | \int \varphi_i^2 (R) e^{-iK_z R_{\alpha z}} dR |^2. \tag{21.2}$$

Here we have dropped the indices of the internal nuclear state and replaced n_i by the normalized probabilities $v_i = n_i N^{-1}$.

We briefly consider two approaches to the calculation of $I(T)$.

(a) The Simplest Model; Direct Calculation. In this model we assume that the vibrations are harmonic and that there are no localized vibrations. After introducing the normal coordinates Eq. (21.2) takes the form

$$I(T) = A|D|^2 F(T) \sum_{\{i_s\}} \exp \left[-\frac{h}{kT} \sum_{\{i_s\}} \omega_s \left(i_s + \frac{1}{2} \right) \right] \tag{21.3}$$

$$\times \left| \int_{\{i_s\}} \prod dq_s\, \varphi_{i_s}^2(q_s) \exp\left(-i\,K_z \sum_s e_{\alpha z,s} q_s\right) \right|^2$$

$$= A\,|D|^2 \prod_s F_s(T) \sum_{i_s} \exp\left[-\frac{\hbar\omega_s}{kT}\left(i_s + \frac{1}{2}\right)\right]$$

$$\times \left| \int dq_s\, \varphi_{i_s}^2(q_s) \exp\left(-iK_z e_{\alpha z,\,s} q_s\right) \right|^2. \tag{21.3}$$

Here we assume that the v_i corresponds to thermal equilibrium, using the notation of section 7, and expand the cartesian shift of the gamma active nucleus in the normal coordinates q_s (see (II.37d) and (II.42); here the index i corresponds to the double index αz)

$$R_{\alpha z} = \sum_s e_{\alpha z,\,s} q_s. \tag{21.4}$$

The coefficients $e_{\alpha z,s}$ are of the order of $N^{-1/2}$ for band vibrations, where N is the number of vibrational degrees of freedom in the whole crystal. Therefore it is convenient to use the series expansion of the exponential

$$\exp(-iK_z e_{\alpha z,\,s} q_s) = 1 - iK_z e_{\alpha z,\,s}\, q_s - \frac{1}{2} K_z^2 e_{\alpha z,\,s}^2\, q_s^2 + \cdots \tag{21.5}$$

Substitution of (21.5) into the matrix element leads to the integral

$$J_{i_s} = \int \varphi_{i_s}^2(q_s)\left(1 - iK_{zs} q_s - \frac{1}{2} iK_{zs}^2 q_s^2\right) dq_s. \tag{21.6}$$

Here we have introduced the parameter $K_z e_{\alpha z,s} \equiv K_{zs}$ to simplify the notation, which for clarity will be interpreted as the projection of the photon wave vector on the s-th normal mode.

Since the square of the harmonic oscillator wave function is an even function, the second term in brackets gives zero upon integration, and taking the normalization into account we find

$$J_{i_s} = 1 - \frac{1}{2} K_{zs}^2 \int q_s^2 \varphi_i{}^s(q_s)\, dq_s. \tag{21.7}$$

To calculate this integral we note that up to a constant factor (whose form depends on the choice of the system of units) q_s^2 represents the potential energy operator and the integral is the quantum-mechanical average potential energy of a harmonic oscillator. Therefore the integral is most simply evaluated using the virial theorem. Since in the Mössbauer effect the α-th gamma-active nucleus is actually displaced, below we set $m_s = m_\alpha$ (the meaning of the mass m_s of a normal mode and the choice of the system of units are discussed in § 7). Thus

$$J_{i_s} = 1 - \frac{1}{2} K_{zs}^2 \bar{q}_s^2 = 1 - \frac{K_{zs}^2}{m_\alpha \omega_s^2} \langle U \rangle_{i_s} = 1 - \frac{1}{2} \frac{K_{zs}^2}{m_\alpha \omega_s^2} E_{i_s}$$

$$= \frac{1}{2} - \frac{K_z^2 e_{\alpha z,s}^2}{2 m_\alpha \omega_s^2} \left(i_s + \frac{1}{2} \right) \hbar \omega_s = 1 - \frac{b_{zs}^2}{2 m_\alpha \omega_s^2 N} \left(i_s + \frac{1}{2} \right) \hbar \omega_s. \quad (21.8)$$

To express the order of smallness of the coefficients $e_{\alpha z,s}$ explicitly we will also use the notation $K_{zs} \equiv K_z e_{\alpha z,s} \equiv b_{zs} N^{-1/2}$, where the b_{zs} are finite (b_{zs} does not depend on N).

Comparison of (21.8) with Eq. (7.12) which gives the analogous matrix element for the band vibrations in the theory of vibronic spectra shows that both expressions are the same, differing only in a factor in the second term which does not depend on the indices i_s. Instead of the factor $\dfrac{a_s^2 m_s}{2N\hbar^2}$ appearing for vibronic transitions, in the Mössbauer effect we have the factor $\dfrac{b_{zs}^2}{2 N m_\alpha \omega_s^2}$.

To obtain the integrated intensity of the Mössbauer line according to (21.3) we take the product of $J_{i_s}^2$ over all normal modes and take a thermal equilibrium average. The dependence of J_{i_s} on the indices i_s over which we average is just as in the corresponding formula for the intensity of the purely electronic line. Therefore the subsequent calculation reduces to a repetition of Eqs. (7.12)–(7.18). The only difference is that instead of the parameters $m_s q_{s0}^2 \hbar^{-2}$ in the intermediate formulas, and the Stokes loss $p_s = \frac{1}{2} m_s \omega_s^2 q_{s0}^2$ in the final formulas, other parameters enter. We write the analog of Eq. (7.17a) in the theory of the Mössbauer effect and attempt to give these parameters a clear physical meaning. We have

$$I(T) = A |D|^2 \exp\left[- \sum_{s=1}^{N} \frac{b_{zs}^2 \hbar}{m_\alpha \omega_s N} \left(\bar{i}_s + \frac{1}{2} \right) \right]$$

$$\equiv A |D|^2 \exp\left[- \sum_{s=1}^{N} \frac{b_{zs}^2}{m_\alpha \omega_s^2 N} k \tau_s \right], \quad (21.9)$$

where $\bar{i}_s$ is the average number of phonons, and τ_s is the "effective temperature" for a harmonic oscillator with a frequency ω_s given by Eq. (7.17b). We rewrite the exponent in (21.9) as the product of two dimensionless factors

$$I(T) = A|D|^2 \exp\left[-\sum_{s=1}^{N} \frac{2R_s}{\hbar\omega_s}\left(\bar{i}_s + \frac{1}{2}\right)\right] \equiv A|D|^2 \exp\left[-\sum_{s=1}^{N} \frac{2R_s}{\hbar\omega_s} \frac{k\tau_s}{\hbar\omega_s}\right],$$

(21.10)

where the parameter

$$R_s \equiv \frac{1}{2} \frac{K_{zs}^2 \hbar^2}{m_\alpha} = \frac{1}{2} \frac{b_{zs}^2 \hbar^2}{N m_\alpha}$$

(21.11)

is dimensionally an energy. The parameter $K_{zs}\hbar$, which has the dimensions of a momentum, is interpreted as the fraction of the recoil momentum transferred to the s-th normal mode by a gamma quantum emitted by the α-th nucleus. Accordingly, it is possible to represent the magnitude of R_s (21.11) as the fraction of recoil energy transferred to the s-th normal mode by a gamma-transition of the α-th atomic nucleus. It is also useful to consider the dimensionless recoil energy r_s

$$r_s \equiv \frac{R_s}{\hbar\omega_s} = \frac{1}{2} \frac{K_z^2 e_{\alpha z, s}^2 \hbar}{m_\alpha \omega_s}.$$

(21.12)

It remains to add that the numerical value of the total recoil energy $R = \frac{E_\gamma^{2*}}{2m_\alpha c^2}$ in systems where the Mössbauer effect is observed lies between 10^{-3} and 10^{-1} eV, i.e., numerically the dimensionless parameter $\sum_s r_s$ is about the same as in vibronic spectra with well-developed purely electronic lines. This means that the relative integrated intensity and temperature dependence of the Mössbauer line are approximately the same as for a purely electronic line.

(b) Quantum-Mechanical Coordinate Distribution of a Gamma-Active Nucleus and the Relative Integrated Intensity of the Mössbauer Line [92, 100]. We return to the basic formula for the relative integrated intensity of the Mössbauer line (21.2) and integrate

---

* $R = \sum_s R_s = \frac{K_z^2 \hbar^2}{2m_\alpha} \sum_s e_{\alpha z, s}^2 = \frac{E_\gamma^2}{2m_\alpha c^2}$, where E_γ is the energy of the gamma

transition (see Eq. (II.43b)).

over all coordinates in the matrix element except the coordinate of the gamma-active nucleus. We find

$$I(T) = A|D|^2 \sum_i v_i \left| \int \rho_i(R_\alpha) e^{-iKR_\alpha} dR_\alpha \right|^2, \qquad (21.13)$$

where

$$\rho_i(R_\alpha) = \int \varphi_i^2(R_1, R_2, \ldots, R_N) dR_1 dR_2 \ldots dR_{\alpha-1} dR_{\alpha+1} \ldots dR_N$$

is the coordinate distribution function of the α-th nucleus, where the lattice is in the i-th quantum vibrational state. The integral in (21.13) is the Fourier transform of the coordinate distribution of the gamma-active nucleus. The separate terms in the sum over i express the contributions of the individual zero-phonon transitions to the overall zero-phonon quasiline — the Mössbauer line. To see the physical meaning of Eq. (21.13) we turn to the particular case of absolute zero. Then the sum over i reduces to the single term

$$I(0) = A|D|^2 \left| \int \rho_0(R_\alpha) e^{-iKR_\alpha} dR_\alpha \right|^2 = A|D|^2 |\tilde{\rho}_0(K)|^2, \qquad (21.14)$$

where $\rho_0(R_\alpha)$ is the coordinate distribution of the α-th nucleus at absolute zero, and $\tilde{\rho}_0(K)$ is its Fourier transform. It follows from the normalization of $\rho_0(R_\alpha)$ that $\tilde{\rho}_0(k=0) = 1$.

Thus the relative integrated intensity of the Mössbauer line at absolute zero is proportional to the absolute square of the K (K is the wave vector of the gamma quantum) Fourier component of the coordinate distribution of the gamma-active nucleus. It is well-known that the distribution of the Fourier transform $\tilde{\rho}_0(k)$ in k space is more smeared-out the more localized the distribution $\rho_0(R_\alpha)$ in coordinate space. The relation between the uncertainties of the coordinate and momentum in quantum mechanics is an expression of this fact. To have a highly probable Mössbauer transition we must have a high probability $\tilde{\rho}_0(k)$ at the point $k=K$. Since for the α-th nucleus in the crystal, $\rho_0(R_\alpha)$ is simply a bell-shaped curve with a single maximum at $R_\alpha = 0$, $\tilde{\rho}_0(k)$ * is a similar function with a maximum at $k=0$.

---

* If the lattice vibrations are harmonic then $\rho_0(R_\alpha)$ is a gaussian function with a maximum at $R_\alpha = 0$, while the Fourier transform of a gaussian function, as is well-known, is another gaussian function with a maximum at $k=0$.

It is thus evident that the Mössbauer line is more intense the lower the absolute value of the photon wave vector (the point K is closer to the origin of coordinates in k space) and the more localized the distribution of $\rho_0(R_\alpha)$ in coordinate space (the distribution $\tilde{\rho}_0(k)$ is more smeared-out and the probability $\tilde{\rho}_0(k)$ drops off less rapidly with $|k|$). In the limit $K \to 0$ $\tilde{\rho}_0(K)$, goes to $\tilde{\rho}_0(0) = 1$, i.e., the relative integrated intensity of the Mössbauer line goes to unity. The distribution $\rho_0(R_\alpha)$ is well localized for small amplitudes of the zero-point vibrations of the α-th atom. We know that with given interatomic force constants the atom is more strongly localized the heavier the α-th atom. The precise form of the distribution $\rho_0(R_\alpha)$ is determined by the localized dynamics of the crystal in the neighborhood of the α-th atom. If the gamma-active nucleus is an impurity atom then the localized dynamics can differ substantially from the dynamics of vibration in the ideal crystal. Localized vibrations in particular can play a large role.

From Eq. (21.14) and our discussion it is also evident that the Mössbauer effect by no means requires an absolutely rigid attachment of the gamma-active nucleus to the crystal. Because of the zero-point vibrations such attachment is not possible even in principle. The role of the crystal environment reduces to a limitation of the motion of the center of mass of the gamma-active nucleus in space. The greater the localization, the greater the probability that the recoil momentum is transmitted to the crystal as a whole and the stronger the zero-phonon Mössbauer line in the center.

From the basic formula (21.13), it is not difficult to understand qualitatively why the relative integrated intensity of the Mössbauer line drops with increasing temperature: with increasing temperature the populations of the higher vibrational levels, which have more smeared-out wave functions in coordinate space, increase.

The probability density of a free atom is distributed uniformly over the whole macroscopic volume accessible to the atom. Its momentum is defined precisely and the zero-phonon line is completely absent from the spectrum. If, however, we limit the spatial region accessible to the atom in any way to microscopic dimensions, then a "zero-phonon" line will appear in the spectrum (to be more precise, a line with no recoil). Thus an increase in the density of a gas containing gamma-active nuclei leads to the appearance of a recoilless line [101]. This line appears in the gamma spectra of ionized atoms placed in a very strong magnetic field which forces the ions to move in circular orbits and limits the motions of their centers of mass perpendicular to the magnetic field [102]. In these directions recoilless lines will also be observed. Diffusional motion of the centers of mass leads to a smearing out of the recoilless line.

The Mössbauer effect is a quantum-mechanical effect since the zero-point vibrations of a crystal are essentially a quantum-mechanical phenomenon.

There is still another possibility for clearly revealing the origin of the Mössbauer line. It is based on extension of the Franck—Condon principle to this effect [97]. A brief treatment of the corresponding version of the Franck—Condon principle is given in Appendix I.

Exercise 25. Proceeding from first-order perturbation theory consider absorption (emission) of a photon by a free stationary atom and find an expression for the recoil energy R of this atom. Estimate R for the yellow line of sodium and for the gamma transition in Ir^{191} (E_γ = 27 keV).

Exercise 26. Estimate the dimensionless total recoil energy $r = (\hbar\bar\omega)^{-1} \sum_s R_s$ for Fe^{57} in metallic iron.

Exercise 27. In what direction does the relative integrated intensity of the Mössbauer line shift if a given gamma-active nucleus is introduced as an impurity atom into a lighter crystal with the condition that the interatomic force constants with the surroundings remain the same.

Exercise 28. Use Eq. (21.10) to study the temperature dependence of the integrated intensity of the Mössbauer line in the low and high temperature limits.

Exercise 29. Estimate the fraction of the whole spectral band in the Mössbauer line with recoil energies R = 10^{-3} and 10^{-1} eV, with T = 0, 100, and 500 K. Assume $\hbar\bar\omega = \hbar\bar{\bar\omega}$ = 0.02 eV.

Exercise 30. (1) Write the explicit form of the distribution $\varrho_0(R_a)$ for a three-dimensional anisotropic harmonic oscillator and calculate its Fourier transform.

(2) Using this Fourier transform illustrate the anisotropy of the Mössbauer effect and the decreasing probability of the Mössbauer line with increasing frequency of the gamma quantum.

§ 22. General Relation between the Mössbauer Effect and Vibronic Spectra

We now consider the general relationship between the Mössbauer effect and vibronic spectra. The results will also serve as a basis for a discussion of the role of localized and pseudolocalized vibrations.

In first-order perturbation theory the vibrational structure defined according to (19.1) corresponds to the matrix element P_{fi}.* Our problem consists in comparing the matrix elements for the two processes. The basic effect of an electronic transition on the lattice oscillators is to change their equilibrium positions, which corresponds in the matrix element to a shift in the origin of one of the vibrational functions in coordinate space. The factor $(-iKR_{\alpha})$ can be interpreted as the recoil momentum given to a nucleus during the emission of a gamma quantum. We will show that it is to be regarded as a displacement of the equilibrium position of an oscillator in momentum space, which appears in the matrix element as a shift of the origin of one of the vibrational functions in momentum space.

Actually, the matrix element in the Mössbauer effect can be represented as follows

$$P_{fi} = \int \varphi_{IIf}(R_1, \ldots R_\alpha, \ldots, R_N) e^{-iKR_\alpha} \varphi_{Ii}(R_1, \ldots, R_\alpha, \ldots$$

$$\ldots, R_N) dR_1 \ldots dR_\alpha \ldots dR_N = \int \eta_{IIf}(p_1, \ldots, \hbar K_\alpha$$

$$-p_\alpha, \ldots, p_N) \eta_{Ii}(p_1, \ldots, p_\alpha, \ldots, p_N) dp_1 \ldots dp_\alpha \ldots dp_N. \quad (22.1)$$

Here K is the wave vector of the gamma quantum, R_α is the position vector of the gamma-active nucleus, the indices I and II refer to the dynamical characteristics of the lattice before and after the gamma transition, and the function η is the Fourier transform of φ. No limitations on the vibrational functions or their changes as

---

* In higher orders of perturbation theory, in the final analysis the specifics of the problem also reduce to the specifics of the matrix elements.

a result of the gamma transition are made. Equation (22.1) is a well-known rule for constructing the Fourier transform of the product of two functions (convolution theorem) [103]. Its validity is not limited to harmonic oscillator wave functions.

A peculiarity of the harmonic oscillator is that its wave function in momentum space is the same as its wave function in coordinate space up to a phase factor (see, for example, [104]). Therefore when the integral on the left-hand side of Eq. (22.1) contains as functions the eigenfunctions of the same harmonic oscillator (or their products) the integrals on the right-hand side turn out to contain products of the same functions of the same oscillator (or system of oscillators). Only the equilibrium position of one of the functions is shifted by the recoil momentum of the gamma quantum. In sum the integral on the right-hand side in (22.1) is simply an overlap integral of the vibrational wave functions which enter into the quantum-mechanical version of the Franck–Condon principle. For a single normal mode we have (see Eqs. (21.3) and (21.6)

$$P_{fi}(s) = \int \varphi_{IIf}(q_s)\, e^{-iK_{zs}q_s}\, \varphi_{Ii}\,(q_s) dq_s$$

$$= \int \eta_{IIf}\,(\hbar K_s - p_s)\, \eta_{Ii}\,(p_s)\, dp_s$$

$$= \mathrm{const} \int \varphi_{IIf}\,(\hbar K_s - p_s)\, \varphi_{Ii}\,(p_s)\, dp_s, \qquad (22.2)$$

where φ_{Ii} are harmonic oscillator wave functions in the coordinate representation (II.52) and η_{Ii} are the same functions in the momentum representation. We have the relation [104]

$$\eta_n\,(p_s) = \frac{1}{\sqrt{2\pi}} \int e^{-ip_s q_s} \varphi_n\,(q_s)\, dq_s = (-i)^n \varphi_n\,(p_s), \qquad (22.3)$$

between them, where n is the number of the vibrational level (n = i) and p_s is the dimensionless momentum of the s-th normal mode.

Thus for a harmonic oscillator we arrive at the known result [26–28] that a shift of momentum, leading to recoil energy in the final formulas, and a shift of coordinate, leading to the appearance of the Stokes loss in the final formulas, are fully equivalent. The harmonic approximation is well justified, hence there is also a good analogy in the direct appearance of vibrational interactions in the structure of the spectrum.

If the vibration of the nucleus is not described in the harmonic approximation, then the Mössbauer effect reduces to the problem of a "Stokes loss in momentum space." This problem is fully equivalent to the usual problem of vibronic transitions without taking the photon recoil energy into account, but in place of the vibrational functions we have the Fourier transforms of the functions which describe the vibrations in coordinate space [96].*

The recoil momentum transmitted to the nucleus in the Mössbauer effect at the time the gamma quantum is emitted can be regarded as a shift of the equilibrium position of an oscillator in momentum space. The usual vibronic transitions in coordinate space can be formulated as a problem of "recoil energy" in momentum space which is equivalent to the usual problem of the Mössbauer effect.[†]

On the basis of the analogy formulated above it becomes quite evident how to apply the Franck–Condon principle to the qualitative interpretation of the Mössbauer effect [97] (see Appendix I).

Exercise 31.   Show that relation (22.3) exists between the coordinate and momentum wave functions of a harmonic oscillator.

§ 23. The Role of Localized and Pseudolocalized Vibrations. Comparison with the Shpol'skii Effect

We can investigate the role of localized vibrations on the basis of Eqs. (21.3) and (21.4). We are evidently not limited to the first terms in the expansion (21.5) since the coefficients $e_{\alpha z \lambda}$ (λ is the localized vibration index) are finite. The mathematical complications arising from this are surmountable (see, for example, [31, 38, 105]). However, we will simply use the results of the

---

* In taking account of the dependence of the electronic matrix element on the vibrational coordinate one must take the Fourier transform of the product of the electronic matrix element and one of the vibrational functions.

† As a rule the interaction with vibrations is more complicated in the vibronic problem, which means that in the equivalent Mössbauer effect problem we need not consider recoil communicated simultaneously to more than one nucleus.

previous section: Eq. (22.2) for the calculation of the matrix element for the Mössbauer effect reduces to a calculation of the matrix element for vibronic transitions. The subsequent operation — averaging over the system of oscillators in thermal equilibrium — is the same in both theories. Equation (22.2), which is valid in the harmonic approximation, relates equally to localized and band vibrations. With this in mind, it is not difficult to see that the calculation of the intensity distribution in the "nuclear-vibrational" Mössbauer band leads to results fully in agreement with the results for vibronic spectra. The only difference is that instead of the Stokes loss and its distribution over normal modes, in the equations for the Mössbauer effect the recoil energy and its distribution over normal modes enters.

Thus we can use all of the equations and discussion of sections 6-17 to describe the vibrational structure of the spectrum and its temperature dependence in the Mössbauer effect. We need only replace the Stokes loss p_s by the recoil energy r_s

$$p_s = \frac{\mathscr{P}_s}{\hbar\omega_s} = \frac{1}{2}\frac{m_s\omega_s^2 q_{s0}^2}{\hbar\omega_s} \rightarrow r_s = \frac{R_s}{\hbar\omega_s} = \frac{1}{2}\frac{K_z^2 e_{\alpha z,s}^2 \hbar^2}{m_\alpha \hbar^2 \omega_s}, \qquad (23.1)$$

where p_s is the dimensionless Stokes loss, and r_s is the dimensionless recoil energy in the s-th normal mode (which could be a localized vibration).

Localized vibrations lead, in the Mössbauer effect as in vibronic spectra, to the appearance of replicas of the zero-phonon line at intervals $m\Omega$, where Ω is the localized vibration frequency and $m = 0, \pm 1, \pm 2, \dots$. Moreover, they affect the integrated intensity and the temperature dependence of the Mössbauer line.

For example, we have for the integrated intensity of the Mössbauer line (the "purely internal nuclear" transition line) in the basic model with L localized vibrations (see Eq. (14.5))

$$S_0^0(T) \sim \exp\left\{-\sum_\varkappa 2r_\varkappa \frac{k\,\tau_\varkappa}{\hbar\omega_\varkappa}\right\} \exp\left\{-\sum_{\lambda=1}^L 2r_\lambda \frac{k\tau_\lambda}{\hbar\Omega_\lambda}\right\} \prod_{\lambda=1}^L I_0\left(\frac{r_\lambda}{\sinh\left(\frac{\hbar\Omega_\lambda}{2kT}\right)}\right), (23.2)$$

while for its replicas at intervals determined by the various combinations of localized vibration frequencies (generated by a set of vibrational quanta $\{\lambda_s\}$ at low temperatures ($kT \ll \hbar\Omega_\lambda$) (see (14.2)-(14.3))

$$S_0(0)_{\lambda_s} = \exp\left\{-\sum_{\varkappa} r_{\varkappa}\right\} \prod_{\{\lambda_s\}} e^{-r_s} \frac{r_s^{\lambda_s}}{\lambda_s!}. \qquad (23.3)$$

Here the index $\varkappa$ refers to band and λ refers to localized vibrations, and r_s is the dimensionless recoil energy per normal mode s according to (23.1). For the details of the remaining notation see section 14.

The role of well-developed pseudolocalized vibrations is the same as that of localized vibrations.

Unfortunately, the experimental methods developed up to the present time do not permit one to measure the vibrational wings in the Mössbauer effect. Therefore the details of the vibrational structure in the wings of the Mössbauer line are only of theoretical interest. We note there that the possibility for experimental investigation of the detailed vibrational structure of the vibronic spectrum is a fundamental advantage over the Mössbauer effect in investigations of the localized vibrational dynamics of a crystal in the neighborhood of an impurity.

In Table 2 we show that the numerical values of the parameters which determine the integrated intensity of the quasilines are nearly equal in the theories of both phenomena. This leads to a similarity of the observed vibrational structure* and temperature dependence of the spectra. We have in mind the "usual" Mössbauer nuclei and not special cases which have very long gamma-transition lifetimes. For optical spectra the recoil energy may be manifested, for example, in spectra involving transitions of free electrons or light excitons in crystals, since it is inversely proportional to the particle mass. In other words, if we have optical transitions involving quasiparticles with a low effective mass we might not be able to neglect the photon momentum despite the fact that it is small in comparison to the quasiparticle momentum.

Thus in addition to the similarity of appearance (astonishingly narrow lines in the spectrum, unusual temperature dependence of the zero-phonon line) and the generality of the underlying causes (specifically the participation of band vibrations in a local-

---

* For the vibrational wings of the Mössbauer effect we speak of the structure to be anticipated in future experiments.

TABLE 2

| | Vibronic transitions at luminescence centers | | Mössbauer effect |
	inorganic crystalline phosphors	organic crystalline phosphors (Shpol'skii effect)	
Stokes loss (change of the vibrational potential energy of the crystal) P, eV	From 10^{-3} (rare-earth ions, ruby, uranyls) up to 1.0 (Tl^+ in KCl)	10^{-2} to 10^{-1}	Negligible
Recoil energy (change of the vibrational kinetic energy of the crystal) R, eV	Negligible	Negligible	10^{-3} to 10^{-1}
Width $\Delta\omega$ of the resonant zero-phonon line	At absolute zero $\Delta\omega = \gamma$ (γ is the intrinsic width of the level), when $T > 0$, $\Delta\omega > \gamma$ and $\Delta\omega(T)$ increases with T and rapidly exceeds γ by several orders of magnitude (due to the scrambling of the normal coordinates or nonadiabaticity).		At absolute zero $\Delta\omega = \gamma$, when $T > 0$, $\Delta\omega(T)$ increases with T, exceeding γ by a small fraction of it.
For a real crystal	At absolute zero $\Delta\omega = \Delta\omega_{st}$ $\approx 10^3$ to $10^4 \gamma$ (the width due to lattice inhomogeneities); with $T > 0$, $\Delta\omega = \Delta\omega_{st} + \Delta\omega(T)$		Practically the same as for an ideal crystal

ized process) the analogy between the Mössbauer effect and vibronic transitions extends, in the basic approximations used to describe lattice vibrations, to the quantitative relations of the theory and the characteristics of the spectra.

If the lattice vibrations are assumed to be anharmonic then a relation between these phenomena persists in the sense that one problem reduces to the other via the Fourier transformation formulated in section 22. The complete correspondence between the formulas (in the sense of the prescription for substitution, (23.1)) ceases to be valid. We will return again to this question in connection with the improved versions of the theory.

Of the differences which are manifested in the more precise treatments, in Table 2 we limit ourselves to the most important and significant in principle: the difference in the observed width of the zero-phonon line. The electronic shells are substantially more sensitive to external influences than atomic nuclei. Therefore even the small inhomogeneities of the internal fields which exist in the best real crystals lead to a spread in the energies of the purely electronic transitions for different impurity centers, which increases the observed line width by a factor of three or four orders of magnitude relative to those which follow from the theory for ideal identical impurity centers. Furthermore, the thermal broadening of the zero-phonon line which appears in the more precise versions of the theory is noticeably present in vibronic spectra. In gamma spectra in which the Mössbauer effect is observed, the effects of crystal inhomogeneities and thermal broadening are small corrections — often vanishingly small.

Chapter 4

Vibronic Spectra in Improved Versions of the Theory

We now turn to improved models. The calculations are substantially more complicated and for clarity one uses the most powerful methods presently known for solving quantum-mechanical problems in solid state physics. The development of these methods goes beyond the scope of this text. We limit ourselves here to a discussion of the results, referring the reader to the original papers for further details. We will consider deviations from the Condon approximation in somewhat more detail. In Appendix III we give several formulas useful in calculating and summing Franck—Condon factors in combinations which arise in the theory of quasiline spectra. Deviation from the Condon approximation will be used to illustrate the use of these formulas.

§ 24. Oscillator Strength of the Electronic Transition Dependent on the Vibrational Displacement in the Crystal (Deviation from the Condon Approximation)

The Condon approximation, which consists of neglecting the dependence of the oscillator strength of the electronic transition on the vibrational displacement, is valid for allowed transitions in a stationary lattice (see sections 4 and 7). Nevertheless it is very important to clarify the role of deviations from this approximation in the theory of vibronic spectra. This is especially important for electronic transitions which are forbidden in the stationary lattice, where the transition arises only because of the violation of this approximation.

The effect of deviations from the Condon approximation on continuous vibronic spectra has been treated in [88, 106, 107]. The basic effect is the appearance of a temperature dependence of the integrated intensity of the absorption and luminescence bands.* For forbidden transitions this dependence is strong and there are cases where it is sharply manifested experimentally in a significant growth of the integrated intensity of an absorption band with temperature [108, 109].

The effect of a vibration–dependent oscillator strength on quasiline vibronic spectra is considered in [110–112].

We will approximate the electronic matrix element by a second–order polynomial in the dimensionless normal coordinates of the crystal.

$$D(x) = d_0 + \sum_s d_s q_s + \sum_{s,r} d_{sr} q_s q_r.$$

$$(24.1)$$

We will consider a change of the equilibrium position of the nuclei during the electronic transition. Changes of the interatomic force constants and anharmonicity of the vibrations will be neglected. This refers to both localized and band vibrations.

The transition probability W_{if} for the set of vibrational quantum numbers $\{i_s\}$ of the system of oscillators to change into the set $\{f_s\}$ is given by

$$W_{if}(T) \sim \langle |J_{fi}(D)|^2 \rangle_T.$$

$$(24.2)$$

Here $\langle \ldots \rangle_T$ denotes averaging over the Boltzmann distribution, and the overlap integrals† (taking the dependence of the electronic matrix element on the vibrational coordinate into account) take the form

---

* In speaking of a luminescence band one must ultimately keep in mind that both the quantum yield and the conditions for exciting the luminescence can change with temperature.

† In this case the integrand contains products of wave functions multiplied by certain coordinate-dependent factors. Usually this factor is a slowly-varying function. Therefore we retain the term "overlap integral."

$$J_{fi}(D) = \int \prod_{s=1}^{N} dq_s \varphi_{f_s}(q_s + q_{s0}) \, \varphi_{i_s}(q_s)$$

$$\times \left(d_0 + \sum_s d_s q_s + \sum_{s,r} d_{sr} q_s q_r \right) = d_0 \prod_{s=1}^{N} J_{f_s i_s}$$

$$+ \sum_s \int dq_s (d_s \varphi_{f_s} \varphi_{i_s} q_s + d_{ss} \varphi_{f_s} \varphi_{i_s} q_s^2) \prod_{\substack{r=1 \\ (r \neq s)}}^{N} J_{f_r i_r}$$

$$+ \sum_{s,r} d_{sr} \int \varphi_{f_s} \varphi_{i_s} q_s dq_s \int \varphi_{f_r} \varphi_{i_r} q_r dq_r \prod_{\substack{t=1 \\ t \neq s \\ t \neq r}}^{N} J_{f_t i_t}, \qquad (24.3)$$

where $J_{f_s i_s}$ is the usual overlap integral for the s-th harmonic oscillator which enters in Eqs. (4.3), (4.4), (13.2), etc.

The overlap integral (24.3) is the sum of four terms, and each term in the sum is the product of N (N is the number of normal coordinates) overlap integrals of independent harmonic oscillators. The first term, proportional to the electronic matrix element d_0, gives the results of the Condon approximation. The remaining three terms express the corrections to the Condon approximation. It is evident that these corrections become important in the particular case $d_0 = 0$ which occurs when the electronic transition is forbidden by symmetry considerations in the equilibrium configuration of the impurity center. The vibrational motion of the lattice distorts the symmetry of the impurity center and its electronic shell so that the selection rules which forbid the electronic transition are no longer valid. The coefficients d_s and d_{sr} (they are far smaller than d_0, so that the series in (24.1) converges rapidly*) express the violation of the selection rules for the electronic transition by the lattice vibrations.

To calculate the integral (24.3) we must find the following integrals for the independent oscillators (the oscillator index is dropped):

$$\int \varphi_f(q + q_0) \, \varphi_i(q) \, dq \equiv J_{fi}, \qquad (24.4a)$$

---

* To be more precise, the series obtained after taking the integral (24.3) converges rapidly.

$$\int \varphi_f (q + q_0) \varphi_i (q) q dq \equiv J_{fi}^{(1)}, \tag{24.4b}$$

$$\int \varphi_f (q + q_0) \varphi_i (q) q^2 dq \equiv J_{fi}^{(2)}. \tag{24.4c}$$

The overlap integral J_{f_i} was calculated for band vibrations in section 6 and for localized vibrations in section 13.

The integrals $J_{f_i}^{(1)}$ and $J_{f_i}^{(2)}$ can be expressed in terms of linear combinations of J_{f_i} and $J_{f\pm 1, i}$. The corresponding formulas are given in Appendix III. According to (III.4) and (III.5)

$$J_{fi}^{(1)} = \frac{1}{q_0} \left(f - i - \frac{q_0^2}{2} \right) J_{fi}, \tag{24.5a}$$

$$J_{fi}^{(2)} = \frac{1}{q_0} \left[\frac{1}{q_0} \left(f - i - \frac{q_0^2}{2} \right) \left(f - i - \frac{q_0^2}{2} + 1 \right) J_{fi} - \sqrt{2(i+1)} \, J_{f,i+1} \right]. \tag{24.5b}$$

To obtain the intensity of a particular quasiline we should take the square of $J_{fi}(D)$ (24.3) and sum the contributions of all transitions $i \to f$ to the given quasiline over final states f and average over initial states i with the Boltzmann factor in accordance with (24.2). In this model contributions to a given quasiline come (with no change of frequency) only from those transitions in which the difference $f_s - i_s$ is fixed and equal to some definite number $k_s = f_s - i_s$. For example, the purely electronic line contains contributions from transitions in which $k_s = f_s - i_s = 0$ for all vibrations s, while the vibrational replicas of the purely electronic lines which arise when $k\lambda$ localized vibration quanta λ are created contribute to transitions $f_s - i_s = 0$ ($k_s = 0$ for all $s \neq \lambda$) and $f_\lambda - i_\lambda = k_\lambda$. This means that in fact only one term is retained from the sum over final states f.

In our model the oscillators are independent; therefore the averaging operation $|J_{fi}(D)|^2$ can be carried out for each oscillator independent of the others. It is still necessary to assume that in squaring $|J_{fi}(D)|$, it makes no sense to retain terms higher than second order in the expansion coefficients (24.1). In sum, the problem of finding the intensity of a certain quasiline reduces to calculating the two sums given in Appendix III (Eqs. (III.1) and (III.2)). The other calculations are simple.

The transition probability $W_{if}(T)$ is given by the following expression, which takes the dependence of the electronic matrix element on the vibrational coordinates into account in the form (24.1)

$$W_{if}(T) = \left\{ \left[d_0 + \sum_s \frac{d_s}{q_{s0}} \left(k_s - \frac{q_{s0}^2}{2} \right) \right]^2 + \right.$$

$$+ 2d_0 \sum_{s,r} \frac{d_{sr}}{q_{s0} q_{r0}} \left(k_s - \frac{q_{s0}^2}{2} \right) \left(k_r - \frac{q_{r0}^2}{2} \right) + 2d_0 \sum_s d_{ss} \left[- k_s + \right.$$

$$\left. + 2\bar{n}_s + 1 - 2y_s I_{k_s+1} (y_s) (I_{k_s}(y_s))^{-1} \right] \right\} \times$$

$$\times \exp \left\{ - \sum_s q_{s0}^2 \left(\bar{n}_s + \frac{1}{2} \right) \right\} \prod_s \left(1 + \frac{1}{\bar{n}_s} \right)^{\frac{k_s}{2}} I_{k_s}(y_s). \tag{24.6}$$

Here $k_s = | f_s - i_s |$ is the number of vibrational quanta which are created or annihilated in the s-th oscillator during the electronic transition, $\bar{n}_s$ is the average number of vibrational quanta in the s-th oscillator in thermal equilibrium, and $I_m(y_s)$ is an m-th order modified Bessel function where $y_s \equiv q_{s0}^2 \sqrt{\bar{n}_s (\bar{n}_s + 1)}$ (for more details see Appendix III).

The expressions obtained contain terms of the zeroth, first, and second orders in the expansion coefficients of (24.1). The zeroth order term is proportional to d_0^2 and gives the result of the Condon approximation.

In order to find the integrated intensity of the purely electronic line one must set all $k_s = 0$ in (24.6). It is then easy to see that if we neglect the natural width, the width of the purely electronic line and the other quasilines is zero in the model under consideration. Their integrated intensities are finite. This means that deviations from the Condon approximation do not lead to a vanishing of the δ-function nature of the purely electronic line and its vibrational replicas. If we consider a line of natural width, then a temperature dependence of the width of the purely electronic line turns out to be a consequence of deviations from the Condon approximation. We will discuss this effect below for a forbidden electronic transition.

For a transition forbidden for the equilibrium positions of the ions entering into the impurity center, $d_0 = 0$ and Eq. (24.7) simplifies to

$$W_{if}(T) = \left[\sum_s \frac{d_s}{q_{s0}} \left(k_s - \frac{q_{s0}^2}{2} \right) \right]^2 \exp \left\{ - \sum_s q_{s0}^2 \left(\bar{n}_s + \frac{1}{2} \right) \right\} \prod_s \left(1 + \frac{1}{\bar{n}_s} \right)^{k_s/2} I_{k_s}(y_s). \tag{24.7}$$

Thus we find for the intensity of the purely electronic line

$$W_{ii}(T) = I(T) = \left(\frac{1}{2}\sum_s d_s q_{s0}\right)^2 \exp\left\{-\sum_s q_{s0}^2\left(\bar{n}_s + \frac{1}{2}\right)\right\}\prod_\lambda I_0(y_\lambda). \quad (24.8)$$

Here we have assumed that $q_{s0}^2 \sim N^{-1}$ and $I_0(y_s) = 1 + 0 (N^{-2})$ for all the band vibrations, and we have used the index λ to enumerate the localized vibrations.

Since the integrated intensity of the whole spectrum increases with increasing temperature* for a forbidden transition, the intensity of the purely electronic line relative to the intensity of the whole spectrum decreases with temperature more rapidly for forbidden transitions than for allowed transitions.

It also follows from Eq. (24.8) that for forbidden transitions the intensity of the zero-phonon line is small not only for large Stokes loss, but also for small Stokes loss. Significant intensity occurs only for intermediate Stokes losses.

The intensities of the vibrational replicas of the zero-phonon line for forbidden transitions are given by Eq. (24.7). An analysis of this equation shows that for large Stokes losses the intensity distribution in the quasilines approaches the distribution in the Condon approximation.

For small Stokes losses the spectrum contains intense one-phonon quasilines. The remaining quasilines, including the purely electronic line, are substantially weaker. It is also possible to have the case where the purely electronic line is almost absent, while its vibrational images (not only the one-phonon images, but the two-phonon images, etc.) are quite intense.

The dependence of the electronic matrix element on the vibrational coordinates leads to an increase in the width of the vibrational sublevels with increasing level number. This is a result of the peculiar effect of vibrational motion on the radiative lifetime of the electronic state: the stronger the lattice vibrates in the vicinity of the impurity the less the equilibrium symmetry of the crystal forbids radiative electronic transitions.†

---

* The zeroth moment of the whole forbidden band increases according to $S_0(T) = \sum_s d_s^2(\bar{n}_s + \frac{1}{2})$ (see section 9 and [106]).

† See [110] for a more detailed treatment; in connection with the selection rules concerned here, see [79, 80].

The large radiation widths of the higher vibrational sublevels mean in turn a thermal increase in the width of the purely electronic line. This increase follows the same law as the increase of the integrated intensity of the whole spectrum with increasing T.

A dependence of the widths of the vibrational sublevels on their number also exists* for allowed transitions but it is very small since in this case the relative change in the electronic transition probability is very small (see, for example, [113]).

Despite the strong thermal decrease of the lifetime for a forbidden transition, it remains more than the lifetime for an allowed transition as a rule. Since in our model the width of a purely electronic line is determined only by the lifetime for radiative electronic transitions, despite the thermal growth its width remains substantially less than the width of a purely electronic line for an allowed transition.

Exercise 32. Write down the dependence of the electronic matrix element on the vibrations (24.1) in cartesian coordinates. Find the relation between the coefficients of the expansions in cartesian and normal coordinates.

Exercise 33. Calculate the Franck–Condon factor for $0 \to k$ transitions in a localized mode considering linear and quadratic dependences of the electronic matrix element on the coordinates.

Exercise 34. Derive Eq. (24.6).

Exercise 35. Describe the distribution of intensity in a series of quasilines involving $0 \to k$ transitions of a localized mode, on the basis of Eq. (24.6) with $d_0 = 0$, and compare the results with the conclusions obtained for such an oscillator in the Condon approximation.

Exercise 36. Show that within the framework of the model of section 24 for the purely electronic line, its vibrational replicas involving localized vibrations have zero width.

---

* Evidently in this case the widths of the higher sublevels can be less than those of the lower ones. Such a possibility is contained in the equation for the zeroth moment of the allowed band S_0 of [106] from which we see that S_0 can decrease with increasing T.

§ 25. Anharmonic Vibrations

Anharmonicity of the vibrations plays a significant role in broad structureless vibronic spectra, and in quasiline spectra it often determines the width and shape of the quasilines.

The anharmonicity of the vibrations has especially important effects on the vibrational replicas of the zero-phonon line. While in the harmonic approximation transitions involving creation (annihilation) of localized vibration quanta give quasilines with the natural width γ, when anharmonicity is considered for the zero-phonon transition the widths of the vibrational replicas of the purely electronic line are greater than γ. This is determined by the anharmonic damping parameters of these oscillators, which are usually three of four orders of magnitude greater than γ. Moreover, anharmonicity of the vibrations leads to a temperature shift of the vibrational replicas of the zero-phonon line [38]. Intrinsic anharmonicity of localized vibrations (and well-developed pseudolocalized vibrations) at nonzero temperatures can lead to the appearance of internal structure of these quasilines and to nonequal spacings between them in a vibrational series. These effects are related to the nonequidistance of the vibrational levels of an anharmonic oscillator, so that the energies of the vibrational transitions which accompany the electronic transition depend on the numbers of both the initial and final vibrational levels. A systematic treatment of intrinsic anharmonicity of localized vibrations, together with anharmonic coupling to the band vibrations, has been given in [38].

Equations which give the intensity, position, and shape of the quasiline with anharmonicity taken into account for arbitrary temperature are quite complicated (see [38, 100, 114]). However, in the most important case of low temperatures, $h\Omega \gg kT$, where the localized oscillators are in the zero-th level* and if we neglect intrinsic anharmonicity of the localized vibrations, these equations simplify substantially. If the area of the spectrum is normalized to unity then the quasilines in the luminescence spectrum corresponding to creation of n_1 localized vibration quanta for the vibration $\lambda = 1$, n_2 localized vibration quanta for the vibration $\lambda = 2$, etc., are described by the following approximate formula [38]

---

* Often $h\Omega/k \gtrsim 300$ K, so that liquid nitrogen temperature satisfies the condition $h\Omega \gg kT$.

$$I(\omega)_{n_1 n_2 \dots} = \frac{W_0}{\pi} \frac{p_1^{n_1} p_2^{n_2} \dots}{n_1! n_2! \dots} \frac{\sum_\lambda n_\lambda \Gamma_\lambda(T) + \gamma}{\left(\omega - \omega_0 + \sum_\lambda n_\lambda \bar{\omega}_\lambda(T)\right)^2 + \left(\sum_\lambda n_\lambda \Gamma_\lambda(T) + \gamma_0\right)^2},$$

(25.1)

where W_0 is the integrated intensity of the purely electronic line, ω_0 is the frequency of the purely electronic transition, and $\bar{\omega}_\lambda(T)$ is the localized oscillator frequency renormalized to take the anharmonicity of the vibrations into account,

$$\bar{\omega}_\lambda(T) = \omega_\lambda + \frac{1}{h} \sum_s W_{\lambda\lambda ss}\left(\bar{n}_s + \frac{1}{2}\right)$$

$$+ P \frac{1}{2h^2} \sum_{s_1 s_2} \left\{ V_{\lambda s_1 s_2}^2 \left(\frac{1 + \bar{n}_{s_1} + \bar{n}_{s_2}}{\omega_\lambda - \omega_{s_1} - \omega_{s_2}} + \frac{2(\bar{n}_{s_1} - \bar{n}_{s_2})}{\omega_\lambda + \omega_{s_1} - \omega_{s_2}}\right.\right.$$

$$\left.\left. - \frac{1 + \bar{n}_{s_1} + \bar{n}_{s_2}}{\omega_\lambda + \omega_{s_1} + \omega_{s_2}}\right) - 2V_{\lambda\lambda s_1} V_{s_1 s_2 s_2} \frac{(2\bar{n}_{s_2} + 1)}{\omega_{s_1}} \right\},$$

(25.2)

$$\Gamma_\lambda(T) \equiv \frac{\pi}{2h^2} \sum_{s_1 s_2} \left\{ V_{\lambda s_1 s_2}^2 [(1 + \bar{n}_{s_1} + \bar{n}_{s_2}) \delta(\omega_\lambda - \omega_{s_1} - \omega_{s_2})\right.$$

$$+ 2\delta(\omega_\lambda + \omega_{s_1} - \omega_{s_2})(\bar{n}_{s_1} - \bar{n}_{s_2})].$$

(25.3)

Here $V_{s_1 s_2 s_3}$ and $W_{s_1 s_2 s_3 s_4}$ are coefficients in the expansion of the third and fourth order anharmonicity operators in the dimensionless normal coordinates, and P means that the sum is taken in the sense of a principal value.

Even at low temperatures $\Gamma_\lambda(T)$ and $\bar{\omega}_\lambda(T)$ depend substantially on T. This happens because a temperature which is low for the localized vibrations can be quite high for the band vibrations on whose parameters Γ_λ and $\bar{\omega}_\lambda$ are dependent. This dependence leads to a substantial thermal shift and broadening of the quasiline with changing T in the low-temperature range $h\Omega_\lambda > kT$. It is important that the widths of the vibrational replicas of the zero-phonon lines already at absolute zero temperature exceed the natural width γ by several orders of magnitude. For an estimate we can take $\Gamma_\lambda(0)/\omega_\lambda \approx 0.01$; however, this value depends essentially on the nature of the vibrational spectrum of the crystal and the position of ω_λ relative to the frequencies in this spectrum.

Equation (25.3) gives a nonzero result only for a localized vibration with a frequency ω_λ not exceeding twice the limiting band vibration frequency. This happens because we have treated the cubic anharmonicity of the lattice vibrations to second order and the fourth-order anharmonicity to first order in perturbation theory. In second order the cubic terms can only give two-phonon decays, and the fourth-order terms give only a level shift in first order and no decay or broadening. Therefore in (25.3) we also have the energy conservation law $\delta(\omega_\lambda - \omega_{s1} - \omega_{s2})$ which eliminates decays of more than two phonons. It is necessary to take account of anharmonic processes to higher orders in the anharmonicity parameters to describe the decay of a high-frequency localized vibration.

If the localized vibration frequency is many times the average band vibration frequency* then only decay into three, four, or even more phonons is possible, and the width of a localized vibrational level due to decay into band vibrations will be small. In this case the basic contribution to the localized vibration level width will come not from decay of the localized vibration, but from processes which can be interpreted as scattering of band vibrations by it (the modulation mechanism of localized vibration broadening [115-120]). These processes always appear in second order for the following fourth-order term in the anharmonicity operator

$$\sum_{ij} W_{\lambda\lambda ij} a_i^+ a_j (a_\lambda^+ a_\lambda - \bar{n}_\lambda),$$ (25.4)

where $W_{\lambda\lambda ij}$ is one of the fourth-order anharmonicity coefficients, k_λ is the number of the localized vibration level, and a_i^+ and a_j are creation and annihilation operators for band vibrations.

The corresponding equation for the width $\Gamma_\lambda^{(n)}(T)$ of the n_λ-th quasiline takes the form

$$\Gamma_\lambda^{(n)}(T) = \frac{\pi}{\hbar^2} n_\lambda^2 \sum_{ij} |W_{\lambda\lambda ij}|^2 (\bar{n}_i + 1) \bar{n}_j \delta(\omega_i - \omega_j).$$ (25.5)

When modulation broadening is taken into account the shape of the quasiline is also described by (25.1) if $n_\lambda \Gamma_\lambda(T)$ is replaced by $\Gamma_\lambda^{(n)}(T)$ from Eq. (25.5).

---

* This occurs, for example, in the real case of the light impurity molecules O_2^-, S_2^-, NO_2^-, and the U center, in alkali halide crystals.

Since these processes do not involve annihilation of localized vibration quanta, but only scattering (creation of one phonon and annihilation of another at a frequency different from the first) of band vibrations by localized vibrations, their probabilities are nonzero even for very high localized vibration frequencies. It should be noted that these processes, in contrast to the decay processes, completely vanish at absolute zero since it is necessary that phonons be present for them to occur, i.e., the band vibrations must be thermally excited. However, even for moderate temperatures ($\hbar\omega_\lambda \gg kT \approx \hbar\bar{\omega}_s$, where $\bar{\omega}_s$ is the average band vibration frequency) they can dominate. In particular, for U centers in KCl the modulation mechanism for broadening of the localized vibration (with a frequency $\omega_\lambda = 570$ cm^{-1}) is already substantial even at 20 K [119, 120].*

If the decay mechanism determines the width of an excited level of a localized oscillator, then at low temperatures ($\hbar\omega_\lambda \gg kT$) according to Eq. (25.1) the width of the quasiline increases linearly with the number n_λ. At higher temperatures the rate of growth of the quasiline width slows down as the level number increases. Moreover, as previously noted, it is possible to have internal structure of a quasiline due to both a change in the elastice constants (see section 26) or internal anharmonicity of the localized level.

If the main contribution to the level broadening comes from the modulation mechanism, then according to (25.5) the width increases with the number n_λ] as a square law. Therefore for high-frequency localized vibrations where decay of an excited state into two or three band phonons ($\hbar\omega_\lambda > 3\hbar\omega_s$) is impossible, one would expect a quadratic increase of the quasiline width with line number.

Neglecting changes in the interaction with the electronic transition, anharmonicity of the vibrations does not effect the position, width, or shape of the purely electronic line. However, the equation for the integrated intensity of this line in the presence of localized vibrations is substantially altered and assumes (neglect-

---

*Here we speak of infrared absorption by localized vibrations, and not of transitions involving a change of the electronic state. But since the theory of the width of a vibrational level is in fact developed in these papers, the results can be extended to vibronic levels.

ing certain unimportant correction terms) the same form as in the absence of localized vibrations [38, 100]

$$I(T) = \exp\left\{-\sum_s p_s(2\bar{n}_s + 1)\right\}.$$ (25.6)

Here s enumerates both band and localized oscillators and p_s is the dimensionless Stokes loss for the s-th oscillator.

The reason that the factor $\Pi I_0 \left(2p_\lambda \sqrt{\bar{n}_\lambda(\bar{n}_\lambda + 1)}\right)$ (λ enumerates the localized vibrations, and $I_0(y)$ is the zeroth order modified Bessel function), which occurs in the equation for the relative integrated intensity of the zero-phonon line, vanishes in the harmonic approximation (14.5) is the same as the one which leads to a broadening of the vibrational images of the zero-phonon line: anharmonic interactions bring about a finite lifetime for the excited states of the localized oscillator. The appearance of the factor $I_0(y)$ which retards the decrease of the purely electronic line intensity with temperature is due to transitions between excited levels of the localized vibrations. If we consider (as is found in the harmonic approximation) that these levels have zero width (or the natural width γ) then $1 \to 1$, $2 \to 2$, ... transitions in the localized oscillator make a finite contribution of zero width (or width γ) to the purely electronic line. In other words, in this case the contributions of the transitions $1 \to 1$, $2 \to 2$, ... add to the $0 \to 0$ contribution in the same narrow spectral region of zero width (or width γ).

In fact anharmonicity gives rise to a substantial (three or four orders of magnitude) increase in the widths of the excited vibrational levels $i_\lambda = 1, 2, \ldots$ of the localized oscillators in comparison to the width γ of the lowest level. Therefore the contributions of the zero-phonon transitions between excited levels of the localized oscillator are distributed over a broader interval (by three or four orders of magnitude) than the width of the purely electronic line and do not contribute to the purely electronic line.

The above treatment is not exact. If it were completely true then the thermal drop in the zero-phonon line intensity would be rapid — it would drop off as the occupation probability of the zeroth vibrational level of the localized oscillator, i.e., proportional to the normalizing factor $F(T)$ ($F(T) = [1 - \exp(-h\Omega/kT)]^{-1}$). A rigorous treatment shows [38] that the transitions $1 \to 1$, $2 \to 2$, ... give an unbroadened line in addition to the broadened line. The

former involves transitions in which the localized vibration con-
serves its energy (which could be anywhere within the limits of the
level width) not approximately, but exactly.

Thus some of the zero-phonon transitions from the excited
vibrational levels give a broadened line and some give a line with
the natural width. As a result, in contrast to zero temperature,
the relative integrated intensity of the zero-phonon line decreases
in comparison to the harmonic approximation (although not so
strongly as when only the $0 \rightarrow 0$ transition is taken into account).
The mathematical expression of this fact is the disappearance of
the Bessel function factor (or product of them) in the equations for
the integrated intensity of the purely electronic line.

If the vibrational background about the purely electronic line
is not too intense the transitions between excited levels of the
localized oscillator can form additional peaks in the spectrum at
the position of the purely electronic line. The widths of these are
of the order of magnitude of the widths of the excited vibrational
levels, i.e., they are much wider than the natural width γ and at
the same time narrower (by a factor of 10 to 100) than the vibrational wings of the spectrum. At the center of this comparatively
broad peak there is a purely electronic line of width γ.

The picture described above is strictly valid in the framework of the model treated in this section. Further improvements
of the theory bring in corrections. We will turn to a discussion of
the shape of the "actually observed" purely electronic line after
discussing all of these corrections.

In concluding this section we stress that all of what has been
said above refers to the quasilines themselves — to the purely
electronic line and its vibrational replicas. In interpreting spectra
it is necessary to forget about phonon wings which arise because
electronic transitions can be accompanied by creation and annihilation of band vibration quanta, in addition to creation of localized
vibration quanta.

The theory developed above can, for example, be applied to
the quasilines which are observed in the low-temperature $KBr-O_2^-$
spectrum shown in Fig. 17.* In the spectra of this system meas-

_____

* We cannot eliminate the possibility, however, that some role is also played here by
rotational motion of the O_2^- molecule. In the case of the NO_2^- impurity in some

ured at liquid nitrogen temperature (Fig. 18), the quasilines are already mixed with phonon wings so that one interprets them as structureless vibronic bands on the localized vibration.

In luminescence spectra of activating molecules in alkali halide crystals, like that shown in Fig. 18, we observe a curious fact: there is a decrease in the half-width of the vibronic band with increasing number, i.e., the half-widths of the bands in Fig. 18 i n c r e a s e from right to left [83, 122, 123]. This is interpreted as arising from anharmonic interaction between localized and band vibrations, which leads to a decrease in the average number of band phonons created during the electronic transition with increasing level number of the localized vibration. In other words, the Stokes loss of a band vibration depends on the degree of excitation of the localized vibration where the vibronic transition occurs [124]. Here it is natural to assume that the band width is not determined by the lifetime of the localized vibration but simply by creation (annihilation) of a certain number of band vibration phonons during the electronic transition. We do not apply Eqs. (25.1) and (25.5), which refer to quasilines, to the band as a whole, but rather similar equations which describe the contour of the structureless absorption and luminescence bands. We can, for example, use the moment formulas (Appendix VI).

§ 26. Force Constant Changes

When the electronic state of the impurity changes, the interatomic force constants which couple the impurity to the surrounding particles of the host crystal also change. These changes, which can be substantial in centers of small radius for large Stokes losses (for example, in crystal phosphors of the KCl-Tl type up to 0.3 to 0.5 f, where f is the force constant, [125, 126]) are usually small for centers having sharp quasiline structure in the spectrum. The degree of deviation of the absorption and luminescence spectra corresponding to the same electronic transition from Levshin's [15] mirror symmetry law can serve as a criterion for a change in f. Quite often the deviations from this law (in the sense of differences of the distances between vibration-

---

alkali halide hosts, it has been firmly established that the fine structure of the low-temperature vibronic quasilines is determined by the rotation of this molecule [121].

al replicas of the purely electronic line in associated absorption and luminescence spectra) are small enough that they cannot be distinguished experimentally.

Nevertheless changes of the interatomic force constants during an electronic transition in an impurity center lead to a consequence of primary importance — they bring about a thermal shift and thermal broadening of the purely electronic line even in the harmonic approximation for the lattice vibrations [127, 39, 95]. Because of the narrowness of the purely electronic line these effects can be clearly manifested even for small values of $\Delta f/f$. We note that these effects are an analog of the corresponding thermal broadening of the Mössbauer line [114, 39, 95] and the shift which, in the Mössbauer effect, is due to a change of nuclear mass during the gamma transition and has been called the second order Doppler shift (see [128]).

As the temperature approaches zero the broadening of the purely electronic line goes to zero. If $\Delta f \ll f$ then for very low temperatures T the width is proportional to T^7 and the shift is $\sim T^4$. At high temperatures the line width is $\sim T^2$ and the shift is $\sim T$.

In the presence of localized (or quite sharply developed pseudolocalized) vibrations changes in the force constants during an electronic transition can also produce internal structure in a purely electronic line if the change $\Delta\Omega$ of the localized vibration frequency is larger than its anharmonic damping constant Γ ($\Delta\Omega > \Gamma$) [72, 105, 38]. The reason for the structure in the line can be understood qualitatively: because of the finite change in the localized oscillator frequency during the electronic transition, the energy of the zero-phonon transition depends on the initial level of the localized oscillator. Here we must also take account of the finiteness of the localized oscillator lifetime. Thus at high temperatures where a substantial number of the localized oscillators are in excited states the form of the purely electronic line becomes complex. It depends substantially on the temperature and on the ratio $\Delta\Omega/\Gamma$.

If $\Delta\Omega \ll \Gamma$ then there is no structure in the purely electronic line. The shape of the purely electronic line in this case, as in the absence of localized vibrations, is close to lorentzian.

In the Mössbauer effect the internal structure of the zero-phonon line arising from changes of the vibration frequency during the gamma transition can evidently not be observed, since the change of the vibration frequency is very small and the condition $\Delta\Omega > \Gamma$ is never fulfilled [39].

The conditions for observing the internal structure of a quasiline are favorable in systems in which either the f r e q u e n c y s h i f t or the i n t e r n a l a n h a r m o n i c i t y (which determines the departure from equidistance between the vibrational levels) is not too small. It is essential that at temperatures where some fraction of the localized oscillators are excited, the thermal broadening of the individual components must not lead to smearing of this structure. This undesirable broadening arises for related but essentially different reasons: the s c r a m b l i n g of the normal coordinates (nondiagonal terms in the transformation matrix of the normal coordinates) and a n h a r m o n i c c o u p l i n g between the localized and band vibrations. We might think that the role of the latter factor is small in systems which have well-isolated localized vibrations, for example, in crystals activated with suitable impurity molecules.

§ 27.   N o n a d i a b a t i c   V i b r o n i c   I n t e r a c t i o n

We place under this heading all effects which arise because the nonadiabaticity operator (1.5), which has been neglected in deriving the adiabatic approximation, can be nonzero.

This interaction also leads to a thermal shift and a thermal drop in the lifetime of the electronic level and thereby to a broadening of the purely electronic line and its vibrational replicas. Nonadiabatic broadening can have a sharply selective character when the energies of certain vibrational sublevels of two (or more) electronic states coincide; here a situation arises similar to that of predissociation in molecules. A calculation for shallow electron traps in semiconductors, whose depths are comparable to the energies of the limiting optical phonons so that it is necessary that nonadiabaticity be considered, is given in [129, 130].

However, even when there are deep localized electronic states in impurity centers of ionic crystals, nonadiabaticity (which is usually a small correction here) can have pronounced effects because of the extreme sensitivity of the purely electronic line and its properties. For small changes of the interatomic force constants during the electronic transition it can be the basic cause of thermal broadening of the purely electronic line.* In particular, it is important when there are other electronic states in the neighborhood of the excited electronic level, for example, if the corresponding electronic level is split into components whose separations are of the order of 10^{-2} eV. Thus in [131] it is shown that, assuming nonadiabatic coupling between the excited levels of Cr^{3+} to be the reason for the thermal shift and broadening of the R line in ruby, one finds conclusions in agreement with experiment.

We cannot exclude the possibility that nonadiabaticity is also the factor which determines the linewidth in spectra involving transitions inside the f shells of rare-earth ions in crystals. If this is so then one would expect quite substantial differences in the widths of the purely electronic lines for different rare-earth activators introduced in the same host, depending on the number and position of the components in the system of f levels.

§ 28. Effects of the Inhomogeneous Structure of the Host Crystal

In treating the theoretical widths of quasilines in the vibronic spectra of crystals containing impurities we should make a clear distinction between two cases: (1) impurity centers completely identical; (2) impurity centers not completely identical due to inhomogeneous structure of the host crystal or inhomogeneous isotopic composition of the impurity center. All of the previous theoretical development has assumed case (1); now we will turn to case (2), which corresponds to the properties of a real set of impurity centers.

(a) Effects of Fields on Point Lattice Defects. From the theory developed for case (1) it follows that the width of the purely electronic line should become of the order of the radiation width, i.e., $\Delta k \approx 10^{-3}$ cm^{-1} or less, as the temperature approaches abso-

---

* But not its vibrational replicas, whose widths are primarily determined by vibrational relaxation in which there is a substantial contribution from adiabatic (i.e., due to the interaction between the vibrations of the system with fixed electronic states) vibrational relaxation. Here we should keep in mind that if we go beyond the framework of the adiabatic approximation then part of the "anharmonicity in a given electronic state" is due to nonadiabaticity.

lute zero.* It is well known experimentally that with decreasing temperature the line narrows and the frequency of its maximum shifts. But the narrowing occurs only down to a certain temperature T_0 equal, for example, to about 50 K [122] for the R_1 line of synthetic rose ruby. Further lowering of the temperature hardly affects the linewidth. The width remains constant with a value 10^3 to 10^4 times the radiation width. For rose ruby at 77 K, $\Delta k = 0.3$ cm^{-1}, and at 2 K, $\Delta K = 0.23$ cm^{-1} [132]. Such a temperature dependence of the width of the purely electronic line is also observed for Yb^{3+} in CaF_2 [133].

This behavior of the purely electronic line at low temperatures can be explained by the effect of inhomogeniety of the host crystal arising from internal strains or point defects. This point of view is supported by the strong dependence of the low-temperature "residual" width on the quality of the crystal and by investigations of the effect of compressing the crystal on the position and splitting of the lines [132].

We will make a simple estimate for point defects and show that the width of the purely electronic line observed at low temperatures (Δk of the order of 0.1 to 0.3 cm^{-1}) can actually be explained by such effects. Internal strains (dislocations) make a contribution of the same order of magnitude.

We assume that the impurities are in slightly different situations due to external electric fields arising from point lattice defects, i.e., other impurity atoms of the same or another kind, and lattice defects of the host crystal. The defects will be regarded as electric dipoles having moments $|D| = D = ed$, where e is the electronic charge, and d is a characteristic lattice dimension, for example, the shortest distance between neighboring sites. The smearing of the levels involved in the purely electronic $I \rightarrow II$ transition will be estimated as the average change of the purely electronic energy $\Delta \varepsilon_0 = \Delta E_{00}$ due to a static dipole field of strength edL^{-3}, where L is of the order of the most probable distance to the nearest point defect. We make the estimate for allowed dipole transitions. We have

$$\Delta \varepsilon_0 \approx \frac{ed}{L^3} e\, (\overline{x}_{II} - \overline{x}_I) \approx \frac{(ed)^2}{(ld)^3} \alpha = \frac{e^2}{d} \frac{\alpha}{l^3}, \qquad (28.1)$$

---

* With the exception of the particular case where one of the combining electronic levels is strongly broadened by nonadiabatic radiationless transitions even at absolute zero.

where we choose L as the number of distances d such that $L = ld$, and α is a coefficient which is the ratio of the change in the average electron orbital radius $\overline{x}_{II} - \overline{x}_I$ in the transition under consideration to d. For a numerical estimate we use $e^2 d^{-1} = 5$ eV and $\alpha = 0.2$. Then $\Delta\varepsilon_0 \approx l^{-3}$ eV or the change in the photon wave number is $\Delta k \approx 10^4 \, l^{-3}$ cm^{-1}, where the factor l^3 can be regarded as the average number of lattice sites per defect. In good crystals with low activator concentrations $l \approx 10^4$ whence $\Delta k \approx 1$ cm^{-1}. Due to lower values of α, Δk can be significantly lower for allowed dipole transitions. For transitions inside a well-screened shell, such as $f - f$ transitions in rare-earth ions, the dipole moment does not change and $\Delta\varepsilon_0$ is substantially weaker and decays faster with increasing interaction distance so that $\Delta\varepsilon_0$ can be substantially less.

The widths of the vibrational replicas of the purely electronic line are of the order of 1 cm^{-1}, as we have shown, for case (1) at low temperatures. Therefore here too the contribution from inhomogeniety is noticeable. It is more important in special cases where the localized vibration can only decay into three (or more) phonons and the vibrational broadening is less.

In Mössbauer lines, of which the purely electronic line is an analog, the radiation width is often observed. The essential difference is that the atomic nucleus is many orders of magnitude less sensitive to lattice inhomogeneity than the electronic shell. However, we might search for experimental methods for eliminating the effect of statistical broadening of the purely electronic line by lattice inhomogeneities and find a purely electronic line with the radiation width. We have the beautiful example where the analogous problem for atoms in a gas has been solved — the Aleksandrov experiment [134] which determines the intrinsic line shape using beats, in which the experimental method allows one to eliminate the Doppler effect. A theoretical treatment by Purga [135] has shown that the use of this method for the spectra of impurities in crystals actually should eliminate the effect of lattice vibrations on the shape of the purely electronic line, and one should see a line of the radiation width. The carrying out of this experiment would be an important step forward in investigations of the optical analog of the Mössbauer line.

(b) The Purely Electronic Line as a Luminescent Indicator of the State of an Impurity in a Crystal. Because it is very narrow, the purely electronic line is quite sensitive to the situation of the impurity in the crystal. Therefore it is a highly sensitive luminescent indicator for solving a number of physico-chemical and crystal-chemical problems. In particular, investigations of the purely electronic line can give new information on the interaction between luminescent centers [136].

Even the "averaged" purely electronic linewidth of 1 cm^{-1} is quite sensitive. Thus in [137] in the low-temperature spectra of luminescent centers in GaP about 30 purely electronic lines were observed which correspond to luminescent centers of identical chemical composition but differing from each other in the distance between the acceptor and donor which compose the center.

The series of papers by Kaplyanskii on piezospectroscopic effects [138] can also be regarded as an example of the use of the purely electronic line to study the fine details of luminescent centers by spectroscopic methods. If one could measure the "true" purely electronic line its width would decrease by three or four orders of magnitude. However, it should be stressed that this does not provide the precision of the Mössbauer effect; the high sensitivity of experiments using this effect provided the first opportunity to observe the s h i f t of the line.

If we were to succeed in extending the method of quantum beats to crystals we would have only solved the problem of determining the s h a p e of the purely electronic line. The theoretical investigations of Purga [135] come to the conclusion that one should see the natural line shape, i.e., the shape due to the finite lifetime of the electronic state.

In Shpol'skii spectra one often sees peculiar "multiplets" — groups of closely spaced lines with frequency differences from several units to tens of cm^{-1}. It is characteristic of these groups that they are replicated over the whole spectrum without a change in the distance between lines, while the group of lines which corresponds to the purely electronic transition is resonant. From this Shpol'skii concluded that the different lines in the group correspond to different spatially distinct impurity molecules with different crystal field conditions, for example, the orientation relative to the crystal axes of the host crystal [22].

Direct experimental confirmation of the correctness of this

point of view has come from coronene in N-heptane by Svishchev, who was able to study the luminescence by selective excitation of individual components of the groups [139].

It also follows from these experimental facts that field in-homogeneities in the host actually can shift the purely electronic line by 1 to 10 cm^{-1}. Comparatively large shifts (up to several tens, or even hundreds, of cm^{-1}) should not be regarded as unlike-ly in large molecules. We incline to think that in Shpol'skii spec-tra each component of the group in turn is broadened by lattice in-homogeneities. The estimate which we have made above applies to this broadening.

§ 29. The Isotope Shift

Isotope shifts of the lattice vibration frequencies appear in the spectra of crystals with impurities as an analog of the corre-sponding shifts in the vibrational spectra of molecules. In vibronic spectra they can occur in two ways: (1) as a shift (broadening) of the purely electronic line or (2) as shifts (broadenings) of the lines in a series of replicas of the purely electronic line at the frequen-cy of one of the localized (pseudolocalized) vibrations. The shift of the purely electronic line arises from a change in the total zero-point energy of the lattice vibrations; the second kind of shift arises from the change in the frequency of one individual localized normal mode or a packet of normal modes forming a pseudolocal-ized vibration. Below we will see that while a shift of the purely electronic line occurs only with a change of the vibration frequency in the electronic transitions, the second kind of shift appears even in the absence of a frequency shift.

(a) Isotope Shift of the Purely Electronic Line. An isotope shift of a purely electronic line has been observed and interpreted in the R$_1$ line of ruby by Schawlow [132]. The vibrational isotope shift of the purely electronic line is a sharp illustration of the fact that all lines in crystal spectra, including the purely electronic line, are essentially v i b r o n i c lines.*

* In the Mössbauer effect the nuclear level scheme itself changes drastically in going from one isotope to another. Because of this in the Mössbauer effect there is no iso-tope shift in the "proper sense" and we lose an additional possibility for studying lattice dynamics.

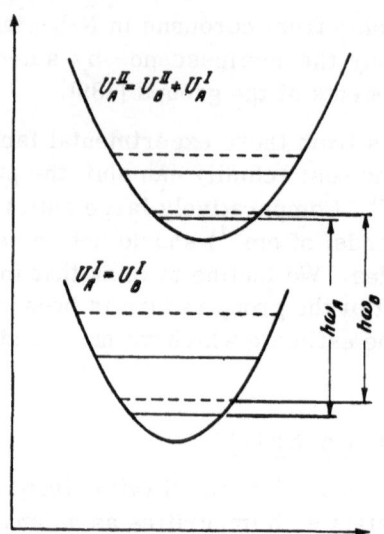

Fig. 23. Isotope shift of a purely electronic line when an A isotope is replaced by a B isotope, equal to $\frac{1}{2}h(\omega_A - \omega_B)$. It arises from the difference in the zero-point vibrational energies since, because of the mass difference, $\omega_I \neq \omega_{II}$. The potential curves do not depend on the kind of isotope, $U_A^I = U_B^I$ and $U_A^{II} = U_B^{II}$. If the potential curves of the ground and excited states are the same ($U_A^I = U_A^{II}$), there is no isotope shift.

We treat the problem in the adiabatic approximation. To clarify the physical picture it is useful to turn to the simplest potential curve diagram (Fig. 23). We consider a purely electronic line at absolute zero where it corresponds to a transition between the vibrational ground states. The forces between ions in the impurity center depend very little on the isotopic composition of the atomic nucleus and we neglect this effect. In other words, the adiabatic potential is regarded as independent of the isotopic composition of the nuclei which enter into the impurity center.* In the language of the simplest diagram this means that the potential curves are completely identical for both isotopes.

---

* This is valid in the adiabatic approximation. If we take account of nonadiabaticity, effects appear which can be reduced to a dependence of the potential curves on the atomic masses.

If there is no change of the interatomic force constants during the electronic transition then there is no isotope shift. With a change of the interatomic force constants there is a contribution from the difference $h\Delta\omega = \frac{1}{2}h(\omega^{II} - \omega^{I})$ of the zero-point vibration energies to the energy of the purely electronic absorption transition. In multicoordinate models this corresponds to $\Delta\omega = \frac{1}{2}(\sum_s \omega_s^{II} - \sum_r \omega_r^{I})$, where the indices I and II denote the electronic states, and r and s are the numbers of the normal coordinates. We let the axis of the normal coordinate system change during the electronic transition. Since with given interatomic force constants the frequency depends on the mass of the vibrating particles, the small contribution $\Delta\omega$ to the purely electronic line frequency is different for different isotopes. This also gives rise to a small isotope shift which, because of the narrowness of the purely electronic line, can be distinctly seen in low-temperature spectra.

The magnitude of the isotope shift $\Delta\omega$ of the purely electronic line arising from replacement of the f-th atom by its isotope is given (assuming that the mass change is small, $(\Delta m_f/m_f \equiv \gamma \ll 1)$ by [140]

$$\Delta\omega = -\frac{\gamma}{4}\left\{\sum_s (e_{sx,II}^2 + e_{sy,II}^2 + e_{sz,II}^2)\,\omega_{0s}^{II} - \sum_r (e_{rx,I}^2 + e_{ry,I}^2 + e_{rz,I}^2)\,\omega_{0r}^{I}\right\},$$
(29.1)

where e_{sx}, e_{sy}, e_{sz} and e_{rx}, e_{ry}, e_{rz} are coefficients of the transformation matrix between the cartesian coordinates of the f-th particle and the normal coordinates (see Appendix II), ω_{0s} is the normal mode frequency for $\gamma = 0$, and the indices I and II refer to the ground and excited electronic states, respectively.

For the R_1 line of ruby the shift arising from replacement of the basic isotope Cr^{52} by Cr^{50} amounts to 0.27 cm^{-1} [132]. From this we can estimate the change of the average chromium vibration frequency in the crystal during the transition into the excited electronic state. It turns out to be a small quantity of the order of 10 cm^{-1} [140]. Therefore we cannot avoid the possibility that non-adiabatic interaction between the electronic and vibrational motion plays a role in the isotope shift of the R_1 line.

(b) Isotope Shifts in Vibrational Series. The problem here reduces to a calculation of the change in the frequency of a localized (or pseudolocalized) vibration due to a change of the mass of one of the atoms in the impurity center. To first order in γ the frequency shift $\Delta\Omega$ of a localized vibration is given by the easily interpreted equation

$$\Delta\Omega = \frac{\gamma}{2}\,(e_{\lambda x}^2 + e_{\lambda y}^2 + e_{\lambda z}^2)\,\Omega, \qquad (29.2)$$

where Ω is the localized vibration frequency and the sum $e_{\lambda x}^2 + e_{\lambda y}^2 + e_{\lambda z}^2$ of the squares of the transformation coefficients between the cartesian displacements of the f-th atom and the localized normal coordinate determines the degree to which the f-th particle participates in the localized vibration (see Appendix II, section 7). Equation (29.2) quantitatively expresses two quite clear dependences. In the first place the isotope shift is greater the greater the relative mass change γ. In the second place with a given value of γ the isotope shift in the spectrum is greater the more the f-th atom participates in the given localized vibration.

Crudely speaking, the role of the f-th atom is determined by the decay law of the localized vibration. Indeed, Eq. (29.2), which takes into account the degree to which the f-th atom participates in the localized vibration, is substantially more exact — in particular it takes account of the symmetry of the impurity center. For example, when the center has inversion symmetry the localized vibrations can be classified as even or odd. In even localized vibrations the central atom (the impurity itself) does not move. Therefore replacement of an impurity atom by an isotope in such a center does not affect the localized vibration frequency.

The isotope shift $\Delta\Omega$ of the localized vibration frequency leads to a difference between the frequencies in the spectra of the isotope i and the isotope i', which increases linearly with the line number n

$$\omega_{i'n} - \omega_{in} = \Delta\omega_{i'i} \pm n\Delta\Omega, \qquad (29.3)$$

where $\Delta\omega_{i'i}$ is the shift of the purely electronic line and n is reckoned from the purely electronic line in the usual way, i.e., ascribing the Stokes replicas of the purely electronic line to positive n (the plus sign corresponds to absorption and minus to emission). $\Delta\Omega$ is the difference between localized frequencies of those vibrations whose quanta are created (or annihilated) in the given transition. For example, for Stokes transitions in the luminescence spectrum, $\Delta\Omega$ represents the difference between the localized frequencies in the electronic ground state.

Anharmonicity leads to violation of the linear relation (29.3).
For a fairly good approximation we can assume that the isotope
shift is a constant fraction of each vibrational quantum which participates in the transition. In this case we have for the isotope shift
between lines

$$\omega_{i'n} - \omega_{in} = \Delta\omega_{i'i} \pm n\Delta\Omega\,(1-\beta n), \qquad (29.4)$$

where β is the anharmonicity parameter. For a crude estimate
we can take $\beta \approx 0.02$. Evidently for high n (n $\approx$ 10) anharmonicity
makes a fully perceptible contribution.

Isotopic lines in the $KBr-O_2^-$ spectrum due to replacement
of the O^{16} isotope by the isotope O^{18} (whose abundance in natural
oxygen is 0.2%) have been observed in [81] and studied experimentally in detail in [141] and theoretically in [140]. Consideration of
anharmonicity turns out to be necessary. Investigation of the isotope shift allowed the author [141] to improve the enumeration of
the lines in the luminescence spectrum and to find the position of
the purely electronic line, which because of large Stokes loss does
not appear in the spectrum. In [142] the isotope shift in $KBr-S_2^-$
due to replacement of the predominant S^{32} isotope by S^{34} was observed.

Thus the effect of the isotopic composition of an impurity
center on the spectral lines of crystals can be manifested in a
braodening and change of the line shape or the appearance of satellite lines. As a rule the satellite lines arise from isotopes of
the impurity itself, and the effects of isotopes in the surroundings
of the impurity reduce to inhomogeneous line broadening. The
isotope shift of a satellite of a fundamental line increases linearly
with n in the harmonic approximation. Anharmonicity disturbs the
linear law but over a small range of n (n $\approx$ 3 or 4) this disturbance
is hard to see. Therefore linearity of the shift in n can serve as
a criterion for the interpretation of a satellite as an isotope shift.
However, the size of the isotope shift for a high-order line (n $\approx$ 10)
already requires the introduction of corrections for anharmonicity.
The largest isotope shifts are to be expected in centers containing
light atoms and for lines with large quantum numbers n. Particularly pronounced effects arise when hydrogen is replaced by deu-
terium (see [143]).

The isotope shift again clearly illustrates the fact that each line in the spectrum of a crystal is essentially a vibronic line. The spectral lines of impurities in crystals are quite sensitive to the effects of isotopes, and not of the light elements only. It is also not necessary that the abundance of the isotope be high — in $KBr-O_2^-$ the abundance is 0.2%.

Exercise 37. Draw the potential curve scheme for a localized vibration and the isotope shift in the spectrum. Take into account a change of frequency during the electronic transition and also consider the anti-Stokes transitions.

Exercise 38. Set up the transition scheme in the presence of two localized vibrations whose frequencies depend on the isotopic mass.

§ 30. The Shape of the Purely Electronic Line. The Nature of the Broadening [144]

The nature of the broadening of a spectral line used for light generation is important in the theory of laser crystals. As we know, a line is regarded as homogeneous if its width is determined by the interactions in each center separately. In this case the observed line is a superposition of lines identical in shape and position.

If, however, the observed spectral line represents a collection of spectral lines arising from spatially distinct and not quite identical centers, in which the center frequencies of the lines differ significantly from one to another in comparison to the width of one individual line, the line is inhomogeneous. Its width is then determined not by the physics of processes at one impurity center, but by the distribution of conditions at the various centers, i.e., by the inhomogeneity of the host crystal. Characteristic peculiarities in the shape of a laser line can arise from nonuniform emptying of the excited states of the various centers during operation of a laser on an inhomogeneous spectral line, for example, the well-known formation of dips ("hole burning") in a laser line.

The nature of the broadening in a spectral line has been explained above. It remains to formulate it in an explicit form and to clarify certain details.

At low temperatures where the purely electronic line already ceases to narrow appreciably with a further cooling of the crystal, its width is determined by the inhomogeneous structure of the host crystal (internal strain, impurities and other point defects, isotopic composition). The purely electronic line is then inhomogeneous.

With increasing temperature the purely electronic line width begins to grow rapidly due to changes in the interatomic force constants or nonadiabaticity. For sufficiently high temperatures the relative contribution of the inhomogeneous broadening becomes small and it can be neglected. Then the purely electronic line can be regarded as homogeneous. The validity of this point of view is shown, in particular, by the development of a crystal laser with an unprecedentedly narrow line. In ruby placed in a special resonator at room temperature, generation of a line of width less than 0.005 cm^{-1} has been observed [145]. For this to happen it is essential that the R_1 line of ruby be homogeneous at room temperature.

The natural line shape is, as is well known, lorentzian. The broadening of the quasiline due to a change in the interatomic force constants, nonadiabaticity, and anharmonicity, also leads to a lorentzian line shape* whose half-width is the sum of the radiation and vibrational half-widths. The radiation half-width can always be neglected in comparison to the latter with the exception of the seldom-realized case of a purely electronic line at extremely low temperatures. The vibrational half-width rapidly increases with increasing temperature. Therefore the high-temperature shape of a quasiline, neglecting the effect of inhomogeneities in the host crystal, is lorentzian.

The low-temperature shape of the purely electronic line is determined by the particular distribution of inhomogeneities in the host crystal and is a sensitive indicator of this distribution. Therefore no universal shape exists. However, quite often the situation is encountered where the broadening is brought about by simultaneous action of a large number of randomly distributed perturbations, such as the distance from the impurity to the nearest dis-

---

* Strictly speaking, if we neglect the radiation width the far wings drop off more rapidly than follows from the Lorentz formula. Moreover, anharmonicity of the vibrations gives rise to a small asymmetry of the shape [38, 98]. The general formula for the asymmetry has been given in [146].

location or point defect, the isotopic composition of the surroundings, etc. If each effect separately does not affect the position of
the purely electronic line too strongly,* then we have a situation
where the resulting distribution is determined by the total effect
of a large number of random variables. It follows from the general results of probability theory that one expects a gaussian distribution for the positions of the purely electronic lines, which
gives a gaussian line shape at absolute zero if we neglect the radiation width.

For finite temperatures, for example at helium temperature,
the vibrational broadening $\Gamma_V(T)$ is already much larger than the
radiation broadening Γ_r and can be comparable to the inhomogeneous broadening. In this case the line shape is the superposition of
lorentzian curves. They have identical half-widths $\Gamma_V(T)$ but their
maxima are located at various frequencies distributed according
to a gaussian law. The resulting shape is described by a curve
known in spectroscopy as the Voigt curve. It was used in astro-
physics to describe the shape of spectral lines already half a cen-
tury ago [147, 148]. There are several ways to construct this
curve approximately but they are evidently not completely satisfactory. Therefore tables and graphs of the Voigt curve have been
prepared for various values of the parameters of its Lorentzian
and gaussian components [149].

In [150] a detailed investigation was made of the shapes of
the purely electronic lines of a number of rare-earth ions in crystals at helium temperatures. In agreement with theory it turned
out that the shape agreed qualitatively with the Voigt curve — the
middle of the line was well approximated by a gaussian curve
while the wings died off much more slowly.

Thus the shape of a low-temperature purely electronic line
is described by the Voigt curve; the greater the parameter which
characterizes its gaussian width the more pronounced is the inhomogeneous nature of this line.

We will now consider the conditions under which we would
expect the narrowest purely electronic line. It is clear from what

---

* If there is a strong effect from one of the factors, for example the different orienta-
tions of the impurity molecules in the crystal or isotope effects in a light impurity,
the line splits into components (see section 28). Our discussion also applies to these
components.

has been said that there are two principal causes of this broadening — inhomogeneity of the host crystal and changes of the interatomic force constants. Both factors have a lesser effect the lower
the c h a n g e in the distribution of electronic density resulting
from the electronic transition of the impurity. The interatomic
forces between the ions or atoms are primarily determined by the
electron density distribution and if it changes little the interatomic
force constants also change little. Inhomogeneity of the host crystal shifts the energy levels of both the ground and excited electronic states of the impurity. Evidently only the d i f f e r e n c e in the
shifts of the terms affects the position of the purely electronic line
in the spectrum, determined again by the difference of the electron densities in the two electronic states of interest in the impurity. In sum we are led to the natural conclusion that the narrowest
lines should be low-temperature purely electronic lines which appear in transitions not involving a change in the electronic configuration, in particular for f-f transitions in rare-earth activators.
In the absence of a frequency change there is also no isotope shift
of the purely electronic line. It is always favorable to choose a
system with homogeneous isotopic composition. We might suppose
that the widths of the f-f transition lines at helium temperatures
are of the order of 0.01 cm^{-1} or even less. Lines with widths of
several hundredths of a cm^{-1} (or less) have been observed experimentally in f-f transition spectra of Dy^{2+} activator ions in CaF_2
[141]. It is possible that the factor which determines the width of
the f-f transition becomes nonadiabaticity even at comparatively
low temperatures. If this is so then we would expect, as noted in
section 27, quite substantial differences in the widths of the purely
electronic lines for different rare-earth activators introduced into
the same host, depending on the number and distribution of the
components in the level system.

Chapter 5

Infrared Absorption and Light Scattering Spectra

§ 31. Infrared Absorption
by Point Lattice Defects

Investigations of infrared absorption have long since become one of the principal sources of information on the vibrations of a crystal. In the last ten years a large number of investigations have been carried out to elucidate the effect of impurities on infrared absorption in various crystals. As a result of experimental [152-159] and theoretical [155, 159-163] investigations, our understanding of the local dynamics of a crystal in the vicinity of an impurity has substantially improved.

Infrared absorption in insulators occurs without excitation of the electronic state, and the spectrum is determined by the vibrational properties (and the dependence of the electronic matrix element on the vibrational displacement) in the electronic ground state of the crystal. In this regard the interpretation is simpler than the interpretation of absorption and luminescence spectra involving electronic transitions.

We will not discuss the general problem of infrared absorption in crystals. We only show how we can obtain equations which describe infrared absorption by an impurity from the general formulas for the vibronic transition probability, and briefly discuss the causes of two-phonon absorption. The reader interested in the general problem of infrared absorption in crystals is referred to the basic survey of Maradudin [159], which contains a discussion of the state of both theory and experiment as of the beginning of 1966. The effects of the interaction of light with elementary excitations in crystals and the appearance of new collective excitations (polaritons) has been treated in connection with infrared absorption in [48], and in connection with scattering and similar problems in [164].

In describing the absorption of light by ideal crystals we should, in addition to energy conservation, also consider conservation of momentum, or more precisely, quasi-momentum. This arises from the translational symmetry of the crystal, which leads to the existence of a phonon quasi-momentum which obeys a conservation law.* If we describe the vibrations of a crystal in the harmonic approximation, which allows only one-phonon transitions, the translational symmetry of the crystal leads to a selection rule which allows infrared absorption only for certain selected frequencies, despite the fact that the normal mode frequencies cover a whole range.

In particular, according to the selection rule in ideal alkali halide crystals in the harmonic approximation it is only possible to have infrared absorption by transverse optical modes of infinite wavelength (vibrations with a wave vector $K = 0$). The spectrum should consist of a δ-function peak. Experimentally this peak corresponds to the long-known strong absorption at the "reststrahl" frequency. (This name arises from the strong reflection of radiation with this frequency from the surface, and its predominance in the spectrum of radiation multiply reflected from the crystal.)

---

* Momentum conservation ultimately is always fulfilled. But in those cases where the mass of the system absorbing the photon is quite large, this law does not lead to any important physical consequences. For example, consideration of the recoil momentum in absorption of an optical photon by an impurity atom gives a shift in the line several orders of magnitude less than the natural width of the line. For an ideal crystal it is important that the translational symmetry creates a quasi-momentum for the elementary excitations in the crystal, including phonons and photons in a crystal, which obeys a conservation law similar to the law of momentum conservation. Thus the quasi-momentum conservation law involves light quasiparticles, and not the crystal as a whole.

The introduction of impurities produces two effects which lead to a quite elaborate pattern of infrared absorption, substantially more interesting and informative than for the ideal crystal.

In the first place, the impurity can lead to the appearance of localized and pseudolocalized vibrations, many of which are active in infrared absorption. (See [159, 160] for the corresponding selection rules.)

In the second place introduction of impurities disrupts the periodicity of the crystal and the selection rules based on conservation of quasi-momentum lose their validity.*

These two effects are ultimately related — they are both a result of a change in the lattice dynamics arising from the introduction of the impurity. One of the most interesting recent observations of details in the impurity infrared absorption spectrum is the appearance of sidebands or phonon wings which arise from the simultaneous action of localized vibrations and band vibrations which have been distorted by the impurity.

Below we give the basic facts about the sidebands and explain how they can be described by equations which are obtained for infrared absorption from the general formulas of section 4.

The extra bands or sidebands were first observed by Fritz [154] in the U-center spectra of a number of alkali halide crystals. He found that the infrared absorption peak of the U center is accompanied on both sides by comparatively weak and broad bands. The maxima of the bands are located at approximately equal distances $\pm \Omega_0$ from the peak, where the frequency Ω_0 is somewhat less than the maximum acoustic vibration frequency of the host crystal. Investigation of the thermal behavior has shown that the intensity ratio of the high and low frequency sidebands varies as $\exp(\hbar\Omega_0/kT)$. From these facts Fritz concluded that the sidebands are a result of transitions in which excitation of the localized vibration of the U center is accompanied by creation or annihilation of a band vibration phonon. The first process gives absorption on the high-frequency side of the peak (Stokes band) and the second gives absorption on the low-frequency side (anti-Stokes band). We

_____

* Moreover, the impurity can induce a dipole moment in a polyatomic homopolar crystal and thus produce infrared-active vibrations in such a crystal (see [159] for more detail).

will show below to what approximation the sidebands can be interpreted as a precise analog of the phonon wings in electronic transitions.

Further investigations have shown that the sidebands can have sharp fine structure, and for this reason provide interesting information about the vibrations of the host crystal.

We now turn to a derivation of the equations which describe the infrared absorption.

We will proceed from the adiabatic approximation and assume that the infrared absorption arises from v i b r o n i c transitions between vibrational sublevels of t h e s a m e electronic state of the impurity. We turn to the equations in part (b) of section 4 and consider the case l = m = 0 in detail, since experiments usually refer to the electronic ground state. The infrared wavelength is tens of thousands of times greater than the dimensions of the impurity center; therefore the phase of the wave within the center can be regarded as independent of the coordinates and the interaction operator between the light and the impurity center can be taken in the dipole approximation (4.2).

We then have for the matrix element of the perturbation operator

$$P_{n'n} = P_{0f,0i} = \int dR \left[\int dr \, \phi_0^* (r, R)\left(-e \sum_i r_i\right) \phi_0(r, R)\right]$$

$$\times \varphi_{0f}(R) \, \varphi_{0i}(R) + \int dR \, \varphi_{0f}(R)\left(e \sum_\alpha Z_\alpha R_\alpha\right) \varphi_{0i}(R) \tag{31.1}$$

The first term on the right-hand side of this equation can be interpreted as transitions between vibrational sublevels due to an electronic dipole moment.

The probabilities of these transitions can be nonzero when the integral over r is nonzero and moreover depends on R. If $| \phi_0(r, R) |^2$ has no center of symmetry as a function of r, then the result of integrating over r and R is in general nonzero.* We can

---

* Strictly speaking, this requirement should be fulfilled for all R. Since in an impurity center, in contrast to diatomic and other simple molecules, there is always vibration which disrupts the inversion symmetry, for some R the integral over r is always nonzero. The situation in principle is completely analogous to the effect of deviations from the Condon approximation on vibronic transitions.

then proceed in the same way as for vibronic transitions involving a change in the electronic state, i.e., we introduce the electronic matrix element $D_{00}(R)$ as a parameter of the theory, and consider the first terms of its expansion in a series in the nuclear coordinates.

We assume that the normal coordinates for the vibrations have already been introduced. Then we can write $D_{00}(q_1 \cdots q_N)$ in the form

$$D_{00}(q_1, ..., q_N) = D_{00} + \sum_s D_s^{(1)} q_s + \frac{1}{2} \sum_{ss'} D_{ss'}^{(2)} q_s q_{s'} + \cdots, \qquad (31.2)$$

where q_s are the normal coordinates. The electronic state indices have been dropped in the expansion coefficients of the matrix elements for simplicity.

It follows from the orthogonality of φ_{0f} and φ_{0i} (they are eigenfunctions of the same Schrödinger equation) that the term D_{00} does not give rise to transitions. If the functions φ_{0f} and φ_{0i} are harmonic oscillator wave functions, then the linear terms in the expansion (31.2) lead to one-phonon transitions, the quadratic ones lead to two-phonon transitions, etc.

In the second term on the right-hand side of (31.1) we have used the dipole moment operator motion of the heavy particles (nuclei) for the perturbation operator. This operator is linear in the cartesian coordinates of the nuclei; therefore it will be linear in the normal coordinates also,

$$e \sum_\alpha Z_\alpha R_\alpha = \sum_s M_s q_s. \qquad (31.3)$$

We can write the matrix element $P_{nn'}$ (31.1) in the form

$$P_{n'n} = \int \left[D_{00} + \sum_s (M_s + D_s^{(1)}) q_s + \frac{1}{2} \sum_{ss'} D_{ss'}^{(2)} q_s q_{s'} + \cdots \right]$$
$$\times \varphi_{0f}(q_1, ..., q_N) \, \varphi_{0i}(q_1, ..., q_N) \, dq_1 \, dq_2 \cdots dq_N. \qquad (31.4)$$

To obtain the infrared absorption probability and its temperature dependence we must take the square of $P_{n'n}$, multiply it by the proportionality coefficient for dipole absorption, and take a thermal average over vibrational states i (see (V.5), (V.7), and (V.9)).

If the lattice vibrations are described in the harmonic approximation, then $\varphi_{0f}(q_1 \ldots, q_N)$ and $\varphi_{0i}(q_1 \ldots, q_N)$ can be represented as products of harmonic oscillator wave functions (2.3). It is not difficult to see that then the terms linear in the normal coordinates in (31.4) give rise only to one-phonon transitions. In other words, neglecting the quadratic term in the expansion of the dipole moment, and assuming the harmonic approximation, we can only have transitions in which the quantum number of a single one of the N oscillators changes by ± 1. This conclusion applies equally to band and localized vibrations. If the functions φ_{0f} and φ_{0i} are calculated taking anharmonicity of the vibrations into account, then already the terms linear in the displacements q_s give two-phonon transitions.

The appearance of sidebands on an impurity infrared absorption line is related to two-phonon processes in which a photon is absorbed, one localized quantum is excited, and one quantum of any band vibration* is excited (or annihilated). We see from (31.4) that there are two reasons for such two-phonon transitions: 1) anharmonic coupling between localized and band vibrations, and 2) bilinear terms in the expansion of the dipole moment operator, i.e., terms of the form $D_{\lambda s} q_\lambda q_s$ where λ is the localized vibration index. Evidently there are no general considerations which allow us to regard one of these causes as always predominant. In general both mechanisms contribute to the appearance of two-phonon transitions. The sizes of the contributions can strongly depend on the host crystal and the impurity.

At present the most completely studied sidebands are those of the U center in alkali halide crystals. In this system we can evidently regard anharmonic coupling between localized and band vibrations [159, 162] as the predominant cause. This is understandable since the vibration amplitude of the light hydrogen impurity ions H⁻ and D⁻ is large so that the anharmonic corrections to the forces should be large. At the same time, in alkali halide crystals the contribution of electronic shell deformation to the dipole moment is small relative to the dipole moment directly connected with lattice vibrations [159].

---

* Strictly speaking, other multiphonon processes can contribute to the sidebands, for example, creation of a localized vibration quantum and two band vibration quanta.

We now consider briefly the problem of interpreting experimental results in which high-frequency localized vibrations enter.

In interpreting impurity spectra generated by high-frequency localized vibrations it is often appropriate to use the simplified but quite clear idea of a "double adiabatic approximation." The essence of this approach is that after separating the electronic motion (fast subsystem) from the vibrational motion (slow subsystem) using the adiabatic approximation, the adiabatic approximation is used again for the vibrational subsystem. This subsystem in turn is separated into two subsystems — the high-frequency localized vibration (fast subsystem) and the band vibrations (slow subsystem).

In introducing this procedure an additional source of error lies in the neglect of the operator for the nonadiabatic interaction between the localized and band vibrations.

Evidently if we use a harmonic description of the vibrations where there is no coupling between the normal modes, the second application of the adiabatic approximation does not introduce any additional error (the nonadiabaticity operator is zero) but is not useful. If we have anharmonic coupling between localized and band vibrations then the nonadiabaticity operator is nonzero but is neglected. Furthermore, in the spirit of the adiabatic approximation it is found that the state of the localized vibration depends on the states of the band vibrations and conversely. In other words, we find a potential surface for the band vibrations for each state of the localized vibration. (In the harmonic approximation these potential surfaces are completely identical.)

Thus the second application of the adiabatic approximation to the vibrations leads to a quite clear picture: we will have a state of the localized vibration which depends parametrically on the configuration of the band vibrations, while for each localized vibrational state in turn there is a potential surface for the band vibrations. The potential surface can be further replaced by an effective potential curve and illustrated conveniently and clearly by a system of potential curves.

Interpretation of the vibrational structure which arises from the band vibrations is now completely analogous to the usual interpretation of the vibrational structure of an electronic transition.

We can, for example, illustrate the potential curves for two levels of the localized vibration between which the transition occurs, and apply the clear ideas of the Franck–Condon principle. We can also use the equations and ideas obtained and developed for electronic transitions in this book.

On the basis of what has been said it is evident that the impurity infrared absorption peaks and their sidebands, treated using a double adiabatic approximation, are completely analogous to the purely electronic line and its phonon wings.

The idea of a double adiabatic approximation is also useful in other problems. For example, in vibronic spectra with sharp structure from a high-frequency vibration such as the luminescence spectrum of $KBr-O_2^-$ (Fig. 18), it is useful to consider a set n = 1, 2, ... of potential curves for the band vibrations, where n is the number of the localized vibration level. In the presence of anharmonic coupling, or scrambling of the localized and band vibrations as a result of the electronic transition, these potential curves are not identical.* In the simplest case their shapes are the same, but there is a shift in the equilibrium position with changing n (for luminescence under conditions where $h\Omega \gg kT$ the question is the shift of the equilibrium position for band vibrations in the vibrational ground state of the localized oscillator in the electronic excited state). In particular, the Stokes losses of the band vibrations depend on n. We also find the dependence of the half-width of a vibronic band on n which is given at the end of section 25 [83, 122, 123].

It is clear that the double adiabatic approximation introduces additional error into the theoretical description. Nevertheless it is useful for qualitative interpretation of experiments and working out new ones, as an approach which gives a quite clear picture of the interaction between band vibrations and localized vibrations. A detailed treatment of the double adiabatic approximation can be found in the work of Sil'd [165].

Exercise 39. Explain the origin of the sidebands and the ratio of the intensities of the Stokes and anti-Stokes bands on

---

* A change of the system of normal coordinates during an electronic transition ("scrambling" of the localized and band vibrations) also leads to a dependence of the potentials on n even in the harmonic approximation, in the form of a new moment during the electronic transition.

the basis of the double adiabatic approximation. (Consider a pair
of effective potential curves of band vibrations for the ground and
excited states of the localized vibration — assume $h\Omega \gg kT$, where
Ω is the localized vibration frequency.

§ 32. Light Scattering by Impurity Centers

Investigations of light scattering have recently received
much stimulation from the introduction and use of laser light
sources in physical laboratories. It has become possible to study
new regions which previously were inaccessible due to the extreme
weakness of the scattering. Experiments on Raman scattering of
light by impurities in crystals also belong to this new area. It may
be inferred that already in the most recent times Raman spectra
have occupied an honored place among the sources of information
about impurity centers and the processes occurring at them (see
[166]).

The lattice vibrations play an even more significant role in
the physics of light scattering by impurities than in absorption and
luminescence. This is because, in addition to determining the vibrational structure in the spectrum of scattered light (vibrational
Raman lines), the vibrations and vibrational interaction cause relaxation which competes with photon emission in an excited electronic state — vibrational relaxation. In principle this relaxation
plays an important role in the physics of the scattering process.
Because of this the secondary radiation of the impurity center does
not consist entirely of scattering but still contains some luminescence. Below we also dwell briefly on the general problem of secondary radiation from an impurity center during optical excitation,
in particular on the classification of secondary radiation.

The theory of light scattering by impurities is constructed
to second order in perturbation theory, using the adiabatic approximation. Although a mathematical description of some of the simplest models is possible by the methods developed in this book,
treatment of good physical models requires development of a
fuller apparatus, which is beyond the scope of this book. Therefore we limit ourselves to a verbal description, citing the original
investigations.

In light scattering by impurities, as in other scattering problems, it is appropriate to distinguish two cases: resonant and non-

resonant scattering. In the first case the frequency of the exciting light lies in an impurity absorption band and in the second case it is not in one.

In principle the situation is simple for nonresonant scattering: there is no absorption, luminescence is not excited, and all of the secondary radiation is scattered light. The theory developed by Placzek [167], in which the vibrational Raman scattering arises from a dependence of the electronic polarizability on the vibrations, is applicable to nonresonant scattering. (See [168, 169] for a refined version of Placzek's theory.)

It is not difficult to show that the spectra have a quasiline structure similar to the structure of ordinary luminescence spectra based on a general qualitative analysis or on a treatment of the nonresonant scattering formula [169].

The zero-phonon line with no change of frequency, the Rayleigh line, is the analog of the purely electronic line. The wings of the Rayleigh line are essentially analogs of the phonon wings in the purely electronic line* which involve acoustic vibrations. The shift of the Raman frequency from the excitation frequency can correspond either to localized or band vibrations.

The lines which correspond to creation or annihilation of localized vibration quanta only (and no band vibration phonons) have a finite intensity. The intensities of the lines involving a change in the number of band vibration phonons by one are infinitesimals of order N^{-1} (N is the number of atoms in the crystal). Since the density of band vibration frequencies is proportional to N (other than at singular points), the total contribution of the band vibrations to the intensity in a finite frequency range is finite. Therefore scattering by band vibrations leads to a more-or-less structureless background or to wings in the Raman spectrum. The peaks in the intensity distribution of the Raman spectrum involve localized (pseudolocalized) vibrations and singularities in the density of vibration frequencies of the host crystal.

Ultimately we have a spectrum whose general character is quite similar to first-order quasiline spectra — to absorption and luminescence spectra. The intensity distribution law is ultimately

---

* The wings of the Rayleigh line can evidently be regarded as Raman frequencies involving acoustic vibrations.

different. From the formulas for the intensity distribution it follows that the probability of multiphonon scattering processes remains small in general even in those centers where the interaction
between the electrons and the vibrations is strong and the Stokes
loss in luminescence is large. As a rule the one-phonon processes
always have the greatest intensity. Two-phonon processes often
give appreciable contributions, but always substantially smaller
than the one-phonon processes.

In describing nonresonant scattering one must consider the
dependence of the matrix element on the vibrational displacement
(i.e., we consider deviations from the Condon approximation) [169].
This is not hard to understand: the principal reason for nonresonant scattering by vibrations is the dependence of the electronic
polarizability on the vibrational displacement, which can affect the
transition probability only through the dependence of the electronic
matrix element on the vibrational coordinates in the adiabatic theory. Incidentally it is evident from this how a factor which appears
as a correction in absorption and luminescence becomes paramount in the scattering theory.

It is evidently promising to use host crystals which have a
first order (one-phonon) vibrational scattering spectrum for experiments on nonresonant scattering by impurities, for example,
activated alkali halide crystals. This circumstance should substantially facilitate the separation of spectra which belong to the
impurities of interest from the remaining components of the spectrum. However, here too defects and impurities which violate the
symmetry of the host crystal and inherently lift the forbiddenness
of one-phonon scattering can substantially complicate the picture.
An especially strong effect arises from an impurity whose absorption resonates with the excitation; due to the resonance it can
luminesce and scatter quite strongly even at a low concentration.
The study of nonresonant scattering by impurities has evidently
become quite possible with the present level of experimental tech-
nique. A theoretical basis for separating the impurity scattering
spectra from the other luminescence can be found in the selection
rules which have been solidly established for nonresonant scattering, including angular and polarization dependences.

In the resonant case, where the excitation frequency is in
resonance with an eigenfrequency of the impurity and quite far

from resonance with the eigenfrequencies of the host crystal, the light acts quite selectively and the properties of the secondary radiation are basically determined by the properties of the impurity center. Therefore it is to be expected that such experiments on resonant secondary radiation will play a large role in investigating impurities. There are always a few systems in which strong first-order scattering from vibrations of the host crystal or a background of scattering and luminescence from the uncontrolled impurities and defects makes experiments on nonresonant scattering by impurities difficult, and it is therefore natural to turn to resonant scattering.

However, resonant excitation involves special experimental difficulties, and complications occur in the theoretical interpretation.

It is well known experimentally that with resonant excitation impurities absorb light and luminesce. In the resonant case the secondary radiation is predominantly luminescence. In this sense the whole history of luminescence investigations is a testimony to the effectiveness of resonant excitation for investigating impurity centers in crystals. But the problem is scattering, or to be more precise, the components of the secondary radiation which do not reduce to ordinary luminescence. In general strong ordinary luminescence makes it difficult to measure this other component which is often weak. The promising experiments, however, are much facilitated if these components fall in a spectral region where there is no luminescence. In particular, it is intuitively obvious that for a system with large Stokes losses, with excitation near the maximum of the absorption band the scattering spectrum can fall on the short-wavelength side of the luminescence band. Part of the scattering spectrum can be located in the vicinity of the purely electronic transition, i.e., in an especially advantageous spectral region, where for large Stokes shifts and low temperatures both absorption (reabsorption) and luminescence are very weak. The theory supports the discussion given above [170, 146] and also leads to an experimentally important conclusion that the scattering spectrum can have pronounced vibrational structure for large Stokes losses also, where the luminescence spectrum is disappointingly close to a gaussian curve.

The theory encompasses the classical problem of resonant

secondary radiation — how to interpret and classify the secondary radiation with resonant photoexcitation. It turns out that the key to solving this problem for impurity centers in a crystal is the correct treatment of the role of vibrations, namely, vibrational relaxation [169, 170, 146]. The idea that relaxation processes ("intermediate transitions") in the excited electronic state play an important role in the classification of secondary luminescence was quite precisely formulated in [171, 172]. The advantage of impurity centers is that the relaxation mechanism is a property of comparatively simple vibrational motion — motion of a set of rather weakly coupled harmonic oscillators — allowing a quite accurate mathematical treatment. This treatment has been given in a number of papers [146, 169, 170, 173-175].

In these papers a version of the general theory of secondary radiation from an impurity center was developed which takes account of vibrations and vibrational relaxation, and the vibrational structure of the spectrum has been calculated for several models and also for particular centers on the basis of the general theory. Below we treat certain general questions, in particular, the problem of classification. We refer the reader to the original papers cited for results on calculation of the general and particular forms of spectra.

Resonant secondary radiation from impurities has been treated in a number of papers [12, 175-182] on the basis of second-order perturbation theory (or equivalent modified approaches).

Effective calculation methods have been developed in these papers and a number of important particular results have been obtained. However, the general pattern of the physics of secondary radiation from impurity centers is not clear. The principal reason for this is the poor understanding of luminescence; the strange conclusion is obtained that the secondary radiation contains no ordinary luminescence. This is not only in clear contradiction with the widest experimental evidence, it is also unsatisfactory from the point of view of theory. Ordinary luminescence is described very well in first-order perturbation theory, but in going to a more precise description of secondary radiation — second-order perturbation theory — the luminescence, which forms more than 99.9% of the total secondary radiation, gets wiped out.

A version of the theory in which an attempt is made to give a regular physical description and classification of the secondary radiation from impurities, including ordinary luminescence, has been worked out in [170, 146, 169]. It turns out, in accord with the simplest physical considerations, that to obtain the correct picture of secondary radiation it is necessary to consider vibrational relaxation in the excited electronic state. This provides a fast "cooling" of the center so that a luminescence spectrum appears which corresponds to the temperature of the crystal. In the first-order theory this important feature is simply accounted for by assuming thermal equilibrium of the lattice vibrations in the initial state. In the second-order theory we do not include the initial conditions for the vibration in the excited electronic state correctly, because we take a state for the initial state in which the impurity is in an unexcited electronic state and is in vibrational thermal equilibrium. The vibrational relaxation in the excited electronic state, which is important for the secondary radiation problem, should be reflected in the model. A mathematical description of this model in second-order perturbation theory should lead to equations containing ordinary luminescence as the most interesting part of the secondary radiation. This program has been realized in [170, 146, 169]. In particular, it was shown that the description of vibrations in the harmonic approximation and the assumption of no vibrational relaxation is a valid approximation in the theory of absorption and luminescence but is completely inapplicable to the theory of secondary radiation. We will not go into details, but only note that the relaxation does not necessarily involve anharmonicity of the vibrations. There is a fine point here; even in the harmonic approximation we can correctly obtain the luminescence if we take account of the fact that the branches of the band vibrations correspond to continuous energy spectra. The dispersion of the normal mode frequencies is reflected in the fact that any localized vibrational excitation of a lattice (in the absence of localized vibrations) decays isoenergetically, even in the harmonic approximation. In this sense relaxation can also occur in the harmonic model. An exact treatment of the problem reduces to analysis of the effect of the spectral composition and the pulse character of the exciting light on the kinetics of the secondary radiation. Here we limit ourselves to remarking that the classification with respect to time characteristics agrees with the classification with regard to spectral properties developed in [170, 146, 169].

Below we will briefly and simply develop the usual physical picture of secondary radiation by an impurity center with resonant photoexcitation, following the results of [170, 146, 169, 175]. This picture is in good agreement with all experimental material, and the simplest reasonable ideas, which are trivial by themselves.

It is well known that the vibrational relaxation time τ_v in an impurity center is several orders of magnitude less than the optical lifetime τ_r ($\tau_v \approx 10^{-12}$ to 10^{-11} sec and for allowed optical transitions $\tau_r \approx 10^{-7}$ to 10^{-8} sec). Therefore the vast majority of the secondary radiation will be emitted after the vibrational relaxation has already occurred, when the impurity center is "cool," and will represent ordinary luminescence. The other part, about a hundredth of a percent of the total secondary radiation intensity, is scattered light (Rayleigh, Brillouin, and Raman scattering) and radiation resembling luminescence but emitted by a still "cooling" center, i.e., during the vibrational relaxation time. The theory of the latter component of the secondary radiation, i.e., the "hot luminescence," has been treated in [166] (a paper by Khizhnyakov, Rebane, and Tekhver).

As a result, we arrive at the following picture. The whole of the secondary radiation spectrum can be represented as a superposition of two spectra: 1) the zero-phonon (Rayleigh) scattering line with the frequency of the exciting light and its vibrational replicas; the frequency of the zero-phonon line does not depend on the properties of the matierial; 2) the zero-phonon (purely electronic) luminescence line and its vibrational replicas; the frequency of this zero-phonon line is determined by the properties of the material (impurity + crystal) and does not depend on the excitation frequency. The vibrational replicas also have phonon wings (replicas involving frequency combinations in which the frequencies of band vibration branches enter).

The spectrum (2) not only contains the whole spectrum of ordinary luminescence, but also a small correction to it — the so-called "hot luminescence" which represents light emission during vibrational relaxation, while the center has still not "cooled" to the temperature of the crystal. This weak luminescence (relative to ordinary luminescence) can be interpreted as luminescence occurring at a high vibrational level which begins to play a role in ordinary luminescence only at high temperatures much higher than the temperature of the crystal. "Hot luminescence" should also arise when the frequency of the incident light is quite high, so that it puts the center in higher vibrational levels than those excited in thermal motion. In practice this means that "hot luminescence" occurs when the excitation is of shorter wavelength than the purely electronic transition. The shorter the excitation wavelength relative to that of the purely electronic line, the "hotter" the vibrational excitation contributing to the "hot luminescence."

We now turn our attention to "hot luminescence" for two reasons. In the first place it is comparable in intensity to the scattering, and can fall in the same spectral region as the Rayleigh scattering. Therefore it must be kept in mind in interpreting scattering spectra. In the second place it is a source of information on processes which occur in a "hot" luminescence center, in particular, on vibrational and radiationless relaxation.

The first attempt to experimentally observe the "hot" luminescence and determine the relaxation time of the localized vibration from its intensity was made for the $KCl-NO_2^-$ system in [183].

We make two remarks on "hot luminescence:"

1) "Hot luminescence" and scattering are not completely independent; there is interference between them. The corresponding corrections to the spectrum were considered in [166].

2) The circle of phenomena associated with "hot luminescence" occurs in dilute gases of molecules under conditions where we observe a dependence of the luminescence spectrum on the frequency of the exciting light (see [184]).

In conclusion we will say some words about the selection rules in the resonant case (see [185]). If we define the scattering according to the classification given above, it is not difficult to conclude that the selection rules should be different than for nonresonant scattering.

Indeed for resonant scattering the Condon approximation gives a nonzero result, and we might think it gives a large contribution to the probability. The contribution to the nonresonant Raman scattering is zero in the Condon approximation. Although deviations from the Condon approximation are more important in resonant scattering than in the theory of ordinary luminescence

[173], the Condon approximation always makes an observable contribution to the resonant scattering. From the results of [146, 170,
169] it is not difficult to see that this contribution to the scattering
should have exactly the same selection rules as ordinary luminescence in the Condon approximation. If we use the selection rules
for the nonresonant region, for example, polarization and angular
criteria, to separate the luminescence from the scattering in the
resonant secondary radiation, there is a risk of losing this "Condon" part of the luminescence, which is scattering in its spectral
composition and physical meaning.

The Franck-Condon Principle

The Franck—Condon principle (FCP) plays a leading role in the theory of vibronic transitions. Despite the fact that the probability of a vibronic transition is fairly easily and precisely calculated by quantum-mechanical methods, the FCP retains its value as a simple way to obtain clear qualitative ideas about the nature of vibronic transitions. For this reason the FCP has also been used in a number of problems connected with the Mössbauer effect.

I.1 The Franck — Condon Principle in the Theory of Vibronic Transitions

In section 4 we considered different formulations of the FCP — classical, semiclassical, and quantum-mechanical versions. In [66] the various versions of the semiclassical FCP were discussed, and they were compared with the probabilities calculated quantum-mechanically. We will give the results of this paper below for the semiclassical version of the FCP which we have formulated in section 4.

Using the Condon approximation, we consider a harmonic oscillator in which only the equilibrium position changes due to the electronic transition.

In Fig. 24 we show the distribution of the vibronic transition probability W_{if} from the initial vibrational levels i = 0 and 1 for a medium Stokes loss ($p = kx_0^2/2h\omega = 7.5$) where k is the force con-

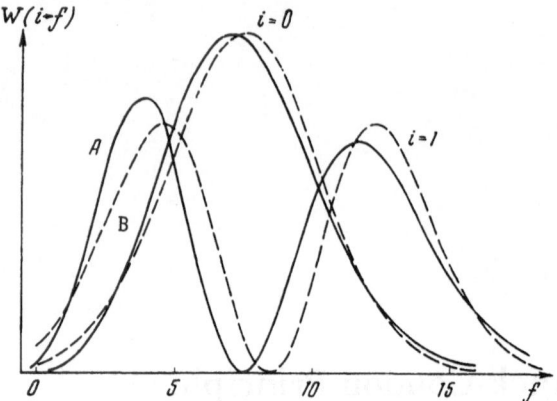

Fig. 24. Distribution of the vibronic transition probability
$W(i \rightarrow f)$ from initial vibrational levels i = 0 and 1 for a
medium Stokes loss (p = 7.5).

stant, x_0 is the distance between the minima of the potential curves,
and ω is the vibration frequency. Curve A is obtained by combining
the smooth curve of points calculated (in the Condon approximation)
as the squares of the corresponding overlap integrals of the vibrational wave functions. Curve B was calculated from the semiclassical FCP formulated in section 4b. It is not difficult to see that
for the simplest model considered here this formulation of the
FCP gives the following analytical expression for the transition
probability

$$W_{if} \sim \varphi_i^2 \left(\frac{p-i-f}{\sqrt{2pk/(\hbar\omega)}} \right), \tag{I.1}$$

where φ_i is the i-th harmonic oscillator eigenfunction.

In Fig. 25 we show the same curves for an oscillator which
has a substantial shift in the equilibrium position during the change
of electronic state (p = 49.5).

It can be seen that the agreement of the quantum-mechanical
and semiclassical results is quite satisfactory. It is better the
greater the Stokes loss. Especially good agreement is found for
the curves of transitions beginning at the zeroth vibrational level.
The relative error of the semiclassical treatment is large at the
ends of the distribution, in particular for the 0−0 transition im-

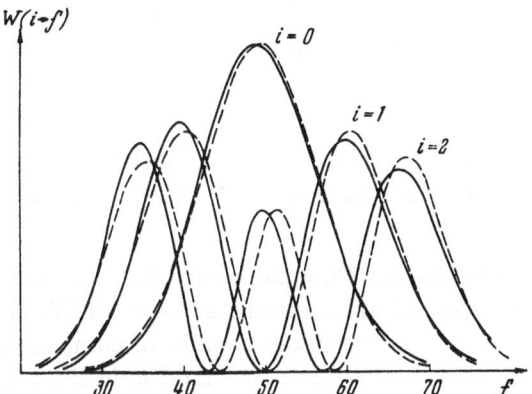

Fig. 25. Distribution of the vibronic transition probability W(i → f) for the initial vibrational levels i = 0, 1,
and 2 for large Stokes loss (p = 49.5).

portant for the formation of the purely electronic line and the
Mössbauer line.

In [186], devoted to a calculation of the vibronic transitions
in a Morse oscillator, a comparison with the semiclassical results
was also given. The agreement between the quantum-mechanical
and semiclassical results is quite good also in the presence of appreciable anharmonicity of the vibrations.

I.2   The Semiclassical Franck – Condon
Principle and the Mössbauer Effect

We will consider the FCP in relation to a number of problems which arise in connection with the Mössbauer effect. We be-
lieve such a treatment to be useful, especially for a deeper understanding of the FCP. It is also curious as still another (and in our
opinion the simplest and clearest) approach to a qualitative understanding of the physics of the interaction with vibrations in the
Mössbauer effect.

In [96] a general treatment of the interaction of a transition
in a fast subsystem (in the electronic shell or inside the nucleus)
with lattice vibrations was given. A general formulation of the
FCP was given in [97] which corresponds to simultaneously considering changes of the adiabatic potential, the nuclear mass, and

the recoil momentum of the photon. In order not to complicate the development with details we turn directly to the basic problem of the Mössbauer effect, i.e., to a transition in which the interaction with vibrations arises only from the photon recoil momentum.

We consider only one harmonic oscillator, which we understand to be a one-coordinate model of the vibrations of an impurity center.

The semiclassical FCP can be formulated in this problem so that the basic physical assumptions are well fulfilled: the internal nuclear state changes extremely rapidly in comparison to the characteristic times for vibrational motion in the lattice.

Evidently only the requirement that the momentum not change at the moment of the transition must be replaced — by the condition that the momentum of the oscillator changes at the instant of the gamma transition by the photon recoil momentum $\pm\hbar K$. It is not difficult to carry out the corresponding graphical construction in the form of the usual potential curve like Fig. 3, which was also done in [97]. This construction, however, is not itself simple: each value of the initial coordinate of the oscillator corresponds to two signs of the momentum. In the graphical construction one must, in determining the lengths of the arrows, choose the change of the momentum $\pm\hbar K$ and express it in terms of the change of kinetic energy.

It is quite simple and clear (and in simplicity and clarity lies all the sense of using the semiclassical FCP) to make a construction which shows instead of the potential curve diagram in coordinate space, i.e., instead of a graph of energy versus coordinate (potential energy), a diagram of kinetic energy in momentum space, i.e., a graph of (kinetic) energy versus momentum. The validity of such a construction is best shown beginning from the quantum-mechanical formula.

The quantum-mechanical distribution of the probability over vibrational sublevels is given in our case (the simplest model) by the expression [see (19.11)]

$$W_{if} = \left| \int dx \varphi_{IIf}^{*}(x) e^{iKx} \varphi_{II}(x)\, dx \right|^{2}, \qquad (I.2)$$

where K is the wave number of the emitted or absorbed photon, and φ_{Ii} and φ_{IIf} are the harmonic oscillator wave functions in the initial and final vibrational states, which are eigenfunctions of the hamiltonian operator

$$\hat{H}_I = -\frac{h^2}{2m}\frac{d^2}{dx^2} + \frac{m\omega^2 x^2}{2}; \quad \hat{H}_{II} = \hat{H}_I + \varepsilon_0. \tag{I.3}$$

Here m and ω are the mass and frequency of the oscillator, and ε_0 is the energy of the "purely nuclear" transition.

In the momentum representation (see section 22)

$$W_{if} = \left| \int dp \eta_{IIf}(p + hK) \, \eta_{Ii}(p) \right|^2, \tag{I.4}$$

where $\eta(p)$ is the harmonic oscillator wave function in the momentum representation. As we know, it is the same (up to a phase factor and the meaning of the parameters) as the wave function of the same oscillator in the coordinate representation [see (22.3)]. The hamiltonian operators whose eigenfunctions are η_{IIf} and η_{Ii} can be written as

$$H_I = -\frac{h^2}{2m_1}\frac{d^2}{dp^2} + \frac{m_1\omega^2}{2}p^2, \quad H_{II} = -\frac{h^2}{2m_1}\frac{d^2}{dp^2} + \frac{m_1\omega^2}{2}(p + hK)^2 + \varepsilon_0, \tag{I.5}$$

where $m_1 = (m\omega^2)^{-1}$.

From Eqs. (I.4) and (I.5) it can be seen that the mathematical problem is completely equivalent to the usual problem of calculating the Franck—Condon integrals. The good applicability of the semiclassical FCP in this problem has been illustrated above. The corresponding graphical construction has been carried out in the form of a "kinetic curve" diagram in Fig. 26. It is fully equivalent to the diagram shown in Fig. 3. The role of the Stokes loss is played by the recoil energy — the "Stokes loss" in momentum space.

We can directly use the conclusions of the comparison of the semiclassical and quantum-mechanical results shown in Figs. 24 and 25. The basic conclusion, which we now state, remains in force: the probability calculated from the semiclassical FCP agrees well with the quantum-mechanical result when there are large recoil energies and for transitions not too far from the most probable one.

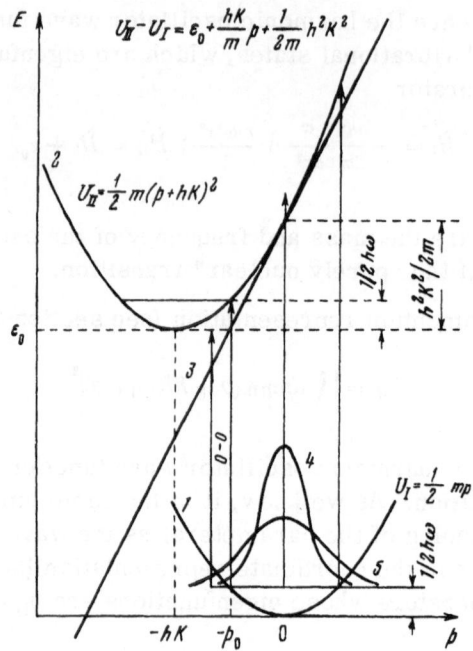

Fig. 26. Kinetic energy curves of the vibrations
$U_I(p)$ and $U_{II}(p) = U_I(p + hK)$ of the initial (curve
1) and final (curve 2) internal nuclear states and
their difference (line 3). The arrows show "internal
nuclear-vibrational" transitions at absolute zero
according to the semiclassical Franck−Condon
principle. Curves 4 and 5 show the quantum-
mechanical momentum probability distributions
in the initial state $p(p)$ with $k\tau = h\omega/2$ (T = 0) and
$k\tau = 2h\omega$. The transition energy is given by the
ordinate of line 3 and the probability of a transition from a state with given p is given by $\rho(p)$.

The integrated intensity of the Mössbauer line or the purely
electronic line is of particular practical importance. From our
discussion of the limitations of the semiclassical FCP formulated
above, it follows that in general one cannot hope for good agreement between the quantitative results of the semiclassical FCP
and the quantum-mechanical results, since the Mössbauer line
either lies far from the most probable transitions (case of large
recoil energy) or occurs at small recoil energies. However, the

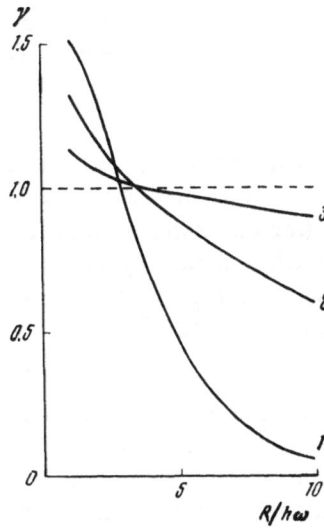

Fig. 27. The ratio γ of the quantum-mechanical and semiclassical zero-phonon transition probabilities as a function of the recoil energy at temperatures kT = 0 (curve 1), kT = hω/2 (curve 2), and kT = hω (curve 3).

general nature of the dependence of the intensity on the recoil energy and the temperature is correctly given.

Actually, the quantum-mechanical and semiclassical zero-phonon transition probabilities for our model are [see (23.2)]

$$I(T)_{qm} = \exp\left(-2\,\frac{R}{h\omega}\frac{k\tau}{h\omega}\right)I_0\left(\frac{R}{h\omega}\frac{1}{\sinh(h\omega/2kT)}\right).$$

$$I(T)_{sc} = \left(4\pi\,\frac{R}{h\omega}\frac{k\tau}{h\omega}\right)^{-1/2}\exp\left(-\frac{R}{4k\tau}\right),\tag{I.6}$$

where I_0 is the zeroth-order modified Bessel function. In Fig. 27 we show the ratio $\gamma = I(T)_{qm}/I(T)_{sc}$ for various temperatures and recoil energies. If the recoil energy is three to five vibrational quanta (i.e., about 0.05 to 0.10 eV with hω = 0.02 eV, which is close to the most frequently encountered recoil energies in the Mössbauer effect), then $I(T)_{qm}$ and $I(T)_{sc}$ roughly agree [97].

From the semiclassical treatment we can easily see that for temperatures $k\tau > (R/2)$, $I(T)_{sc}$ decreases with increasing temperature due to an increase in the smearing of the probability distribution (since the second central moment $\overline{S}_2 = 2Rk\tau$).

In Fig. 26 we give the corresponding graphical construction in which we see with particular clarity that $I(T)_{sc}$ is determined by the probability that the system remain with momentum p_0 defined from the condition $U_{II}(p) - U_I(p) = \varepsilon_0 + (hK/2m)(hK + 2p) = \varepsilon_0$, i.e., $p_0 = -hK/2$.

The temperature dependence is also evident from this — at finite hK (this occurs for localized vibrations) $I(T)_{sc}$ initially increases and then (with temperature $k\tau > R/2$) begins to decrease monotonically; with infinitesimal hK (for one band normal mode $hK \sim N^{-1/2}$, where N is the number of vibrational degrees of freedom) $I(T)_{sc}$ decreases monotonically with increasing temperature beginning at absolute zero. Such behavior is confirmed by an exact quantum-mechanical treatment.

It follows from our discussion that the Franck—Condon principle can actually be used for qualitative interpretation of the Mössbauer effect. The basic features of the Mössbauer effect can easily be understood if we supplement the Franck—Condon principle with an analysis of the causes of the energy width of the lowest vibrational level in the system. It is also useful for a crude estimate of the integrated intensity of the Mössbauer line.

The use of the clear semiclassical Franck—Condon principle is especially advantageous in conjunction with a quasimolecular model of the lattice vibrations in the vicinity of a nucleus which undergoes a gamma transition, similar to those used in the theory of vibronic transitions in impurity centers [187-190, 53, 54]. Evidently the FCP can also be formulated for other phenomena analogous to the vibronic transitions and the Mössbauer effect. In particular it is not difficult to extend the treatment to absorption of neutrinos or neutrons by the atomic nuclei in a crystal.

Appendix II

Some Concepts from the Theory
of Small Vibrations

We will consider small vibrations both in their classical and quantum-mechanical aspects. It is difficult to avoid some repetition here, but the author believes this will lead to a fuller illumination of the subject.

Initially we will be concerned with the classical mechanics of small vibrations of a system of material points about their equilibrium positions. We will see that such vibrations can be described as the vibration of a set of independent harmonic oscillators. In order to obtain the quantum-mechanical description of the vibrations which we ultimately desire, we treat the motion of each oscillator quantum-mechanically. Since we speak of independent harmonic oscillators this is trivial.

In section II.5 we will formulate the normal coordinate problem at the very outset as a quantum-mechanical problem of separation of variables in the Schrödinger equation which describes the small vibrations.

II.1 Small Vibrations of a System
of Material Points

We expand the potential energy of a crystal (molecule) which has $N = 3n^*$ vibrational degrees of freedom in a Taylor series

---

* More exactly, $N = 3n - 6$. See [56] for a treatment of the separation of the three

about the stable equilibrium position of the system and retain terms through the quadratic ones

$$U(x_1, x_2, \ldots x_N) \approx U(x_{10}, x_{20}, \ldots, x_{N0}) + \frac{1}{2} \sum_i \sum_k A_{ik}(x_i - x_{i0})(x_k - x_{k0}).$$
(II.1)

Here the indices i and k denote one of the three cartesian displacements of the L-th particle (L = 1, 2, ..., n) and assume values from 1 to N, x_{i0} is the equilibrium value of the i-th coordinate, and

$$A_{ik} \equiv \left(\frac{\partial^2 U}{\partial x_i \partial x_k}\right)_{x_i = x_{i0}; \ x_k = x_{k0}}$$

It is convenient to choose the equilibrium position x_{i0} as the origin of coordinates x_i. Then

$$U(x_1, x_2, \ldots, x_N) = U_0 + \frac{1}{2} \sum_i \sum_k A_{ik} x_i x_k,$$
(II.1a)

and we find for the l-th component of the force

$$F_l = -\frac{\partial U}{\partial x_l} = -\frac{1}{2}\left(\sum_k A_{lk} x_k + \sum_i A_{il} x_i\right) = -\sum_k A_{lk} x_k.$$

Here we have used the symmetry $A_{ik} = A_{ki}$. From this we obtain N equations of motion

$$m_l \ddot{x}_l + \sum_{k=1}^{N} A_{lk} x_k = 0, \ l = 1, 2, \ldots, N.$$
(II.2)

or in expanded form

$$m_1 \ddot{x}_1 + A_{11} x_1 + A_{12} x_2 + \ldots + A_{1N} x_N = 0,$$
$$m_2 \ddot{x}_2 + A_{21} x_1 + A_{22} x_2 + \ldots + A_{2N} x_N = 0,$$
$$\cdot \ \cdot \ \cdot \ \cdot \ \cdot \ \cdot \ \cdot \ \cdot \ \cdot \ \cdot \ \cdot \ \cdot \ \cdot \ \cdot \ \cdot$$
$$m_N \ddot{x}_N + A_{N1} x_1 + A_{N2} x_2 + \ldots + A_{NN} x_N = 0.$$

We should keep in mind that, because of the system of indices used, the three masses m_1 referring to the same particle are the same, since $m_{Lx} = m_{Ly} = m_{Lz}$.

The well-known method for solving this system, which is a second-order system of homogeneous linear differential equations, is the following. We seek a solution in the form

degrees of freedom associated with translation of the center of mass and the three degrees of freedom associated with rotation of the crystal (molecule) as a whole.

$$\dot{x}_k = a_k m_k^{-1/2} e^{i\omega t},$$ (II.3)

and after substituting this in (II.2) we obtain a system of linear algebraic [equations for the N constants a_k

$$-\omega^2 a_l + \sum_{k=1}^{N} \frac{A_{lk}}{\sqrt{m_l m_k}} a_k = \sum_{k=1}^{N} \left(\frac{A_{lk}}{\sqrt{m_l m_k}} - \delta_{lk} \omega^2 \right) a_k = 0.$$ (II.4)

In expanded form we have

$$\left(\frac{A_{11}}{m_1} - \omega^2 \right) a_1 + \frac{A_{12}}{\sqrt{m_1 m_2}} a_2 + \dots + \frac{A_{1N}}{\sqrt{m_1 m_N}} a_N = 0,$$

$$\frac{A_{21}}{\sqrt{m_2 m_1}} a_1 + \left(\frac{A_{22}}{m_2} - \omega^2 \right) a_2 + \dots + \frac{A_{2N}}{\sqrt{m_2 m_N}} a_N = 0.$$ (II.4a)

$$\dots \dots \dots \dots \dots \dots \dots \dots \dots \dots \dots$$

$$\frac{A_{N2}}{\sqrt{m_N m_1}} a_1 + \frac{A_{N1}}{\sqrt{m_N m_2}} a_2 + \dots + \left(\frac{A_{NN}}{m_N} - \omega^2 \right) a_N = 0.$$

Since the equations are homogeneous the existence of a nontrivial solution* requires that the determinant of system (II.4a) vanish.

We thus obtain the characteristic (or secular) equation

$$\begin{vmatrix} \dfrac{A_{11}}{m_1} - \omega^2 & \dfrac{A_{12}}{\sqrt{m_1 m_2}} & \cdots & \dfrac{A_{1N}}{\sqrt{m_1 m_N}} \\[2mm] \dfrac{A_{21}}{\sqrt{m_2 m_1}} & \dfrac{A_{22}}{m_2} - \omega^2 & \cdots & \dfrac{A_{2N}}{\sqrt{m_2 m_N}} \\[2mm] \cdots & \cdots & \cdots & \cdots \\[2mm] \dfrac{A_{N1}}{\sqrt{m_N m_1}} & \dfrac{A_{N3}}{\sqrt{m_N m_2}} & \cdots & \dfrac{A_{NN}}{m_N} - \omega^2 \end{vmatrix} = 0.$$ (II.5)

This equation, which is an algebraic equation of N-th degree in ω^2, determines the vibrational eigenfrequencies ω_j.[†] It has N roots ω_j^2 (j = 1, 2, ... , N) some of which can be repeated. We assume for simplicity that all N of the roots are distinct.[‡]

* The trivial solution is a_k = 0 (k = 1, 2, ... , N).

† See [191-193] for methods of solving the characteristic equation and other contemporary methods in the theory of molecular vibrations.

‡ Below we will basically follow the development given by V. G. Nevzglyadov in his book [194]. There will be a difference in that we consider the problem in cartesian coordinates, so that in our problem the kinetic energy is already diagonal. If we set a_{ik} = $a_i \delta_{ik}$ = $m_i \delta_{ik}$ and e_{ik} = a_{ik} in Eqs. (98.8) etc. of [194], then (97.10) coincides with our Eq. (II.2).

We substitute one of the roots ω_j^2 into system (II.4). We know from linear algebra that for the simple roots (roots of unit multiplicity) one of the N equations (II.4) is now a consequence of the other equations [195]. We thus have N−1 independent equations for the N constants a_k. We can express N−1 of the constants in terms of one of the N constants, whose value remains arbitrary. We use a_N as the arbitrary constant, which is equivalent to dropping the last equation in the system (II.4a) and eliminating a_N from the remaining equations. We thus obtain a system of N−1 i n h o - g e n e o u s linear equations for the N−1 remaining constants $a_k' \equiv a_k/a_N$

$$\sum_{k=1}^{N-1}\left(\frac{A_{lk}}{\sqrt{m_l m_k}} - \delta_{lk}\omega^2\right)a_k' = -\frac{A_{lN}}{\sqrt{m_l m_N}}, \quad l = 1, 2, \ldots, N-1. \quad \text{(II.6)}$$

The determinant of this system is not zero, since the root ω_j^2 is not repeated. The solution of the system is given by

$$a_k' = \frac{\Delta_k(\omega_j^2)}{\Delta_N(\omega_j^2)}, \quad k = 1, 2, \ldots, N-1, \quad \text{(II.7)}$$

where $\Delta_N(\omega_j^2)$ is the determinant of the system (II.6), i.e., the determinant obtained by suppressing the last row and column in the determinant (II.5), and $\Delta_k(\omega_j^2)$ is the determinant obtained by replacing the coefficients in the k-th column of the determinant $\Delta_N(\omega_j^2)$ by the constant terms in the system (II.6). $\Delta_k(\omega_j^2)$ can also be obtained by suppressing the last row and the k-th column in the determinant (II.5) and multiplying by $(-1)^{N-k+1}$.

Turning to the constants a_k, we can write the result in the following form

$$a_k = b\Delta_k(\omega_j^2), \quad k = 1, 2, \ldots, N, \quad \text{(II.8)}$$

where the arbitrary constant b is defined by

$$b = \frac{a_N}{\Delta_N(\omega_j^2)}.$$

Since ω_j^2 is a root of the secular equation, $\pm\omega_j$ are also roots of it. Therefore turning to the functions (II.3) and substituting the constants a_k from (II.8) into them, we find a particular solution of the equations of motion (II.2) for each root ω_j^2 in the following form

$$x_k^{(j)}(t) = (b_j e^{i\omega_j t} + b_j' e^{-i\omega_j t}) m_k^{-1/2} \Delta_k(\omega_j^2), \quad k = 1, 2, \ldots, N, \quad \text{(II.9)}$$

where b_j and b_j' are two arbitrary constants. Evidently we will have small vibrations if the ω_j are real, i.e., if $\omega_j^2 > 0$. The determinant $\Delta_k(\omega_j^2)$ is a real number since all of the elements in the determinant (II.5) are real. Therefore we find real values of the functions $x_k(t)$ for all t if we choose the constants b_j and b_j' to be complex conjugates, $b_j^* = b_j'$. Setting $b_j = \frac{1}{2} c_j e^{i\alpha_j}$ we can write (II.9) in the form

$$x_k^{(j)}(t) = \Delta_k(\omega_j^2) m_k^{-1/2} c_j \cos(\omega_j t + \alpha_j), \quad k = 1, 2, \ldots, N. \quad \text{(II.10)}$$

The particular solutions (II.9) of the equations of motion (II.2) can be found for e a c h root ω_j^2; therefore we can set up a sum of N particular solutions (II.10) which will also be a solution of the linear equations of motion (II.2). Thus we find

$$m_k^{1/2} x_k(t) = \sum_{j=1}^{N} \Delta_k(\omega_j^2) c_j \cos(\omega_j t + \alpha_j), \quad k = 1, 2, \ldots, N. \quad \text{(II.11)}$$

The functions (II.11) contain $2N$ arbitrary independent real constants c_j and α_j and therefore are general integrals of the equations of motion (II.2), i.e., general solutions for small vibrations of the system of n material points. The integration constants c_j and α_j are determined from the initial conditions.

We note that the quantities $m_k^{1/2} x_k$ are also called reduced displacements.

II.2 Eigenfrequencies, Normal Modes, and Normal Coordinates

We will discuss the general solution (II.11). Each of the coordinates $x_k(t)$ (which describes one of the cartesian displacements of one of the atoms in the crystal from its equilibrium position) is generally involved in a complicated vibrational process which is a superposition of N harmonic vibrations with frequencies ω_j. The roots ω of the secular equation (II.5) are the eigenfrequencies of the small vibrations.

The vibrational behavior of $x_k(t)$ depends ultimately on the integration constants c_j and α_j. This reflects the dependence of the vibrational behavior on the initial conditions. In particular, we could imagine initial conditions in which $c_j = \delta_{js} c_s$, i.e., all of the c_j are zero except c_s. With this choice of initial conditions the coordinate $x_k(t)$ vibrates in simple harmonic motion with a small oscillation eigenfrequency ω_s

$$m_k^{1/2} x_k(t) = \Delta_k(\omega_s^2) c_s \cos(\omega_s t + \alpha_s) \equiv \Delta_k(\omega_s^2) q_s(t), \qquad (\text{II}.12)$$

where the function $q_s(t)$ is called a normal coordinate and describes the vibrations of a harmonic oscillator called a normal oscillator. In all there are N normal coordinates

$$q_j(t) = c_j \cos(\omega_j t + \alpha_j), \quad j = 1, 2, \ldots, N \qquad (\text{II}.13)$$

Using the normal coordinates, the general solution (II.11) can be written in the form

$$m_k^{1/2} x_k(t) = \sum_{j=1}^{N} \Delta_k(\omega_j^2) q_j(t). \qquad (\text{II}.14)$$

We can see that the cartesian coordinates are linear combinations of the normal coordinates. It is important that the transformation coefficients $\Delta_k(\omega_j^2)$ do not depend on time.

The normal coordinates have the important property that the kinetic and potential energies of the system are the sums of squares when expressed in terms of them. We will show that this is so.

The potential energy (II.1), expressed in terms of the normal coordinates, has the form

$$V = U - U_0 = \frac{1}{2} \sum_{i,k=1}^{N} A_{ik} x_i x_k = \frac{1}{2} \sum_{j,l=1}^{N} B_{jl} q_j q_l, \qquad (\text{II}.15)$$

according to (II.14), where

$$B_{jl} \equiv \sum_{i,k=1}^{N} \frac{A_{ik}}{\sqrt{m_k m_i}} \Delta_k(\omega_j^2) \Delta_i(\omega_l^2). \qquad (\text{II}.15a)$$

The kinetic energy is

$$T = \frac{1}{2} \sum_{k} m_k \dot{x}_k^2 = \frac{1}{2} \sum_{j,l} D_{jl} \dot{q}_j \dot{q}_l, \qquad (\text{II}.16)$$

where

$$D_{jl} \equiv \sum_k \Delta_k(\omega_j^2)\, \Delta_k(\omega_l^2).$$

(II.16a)

We must show that all of the coefficients B_{jl} and D_{jl} with $j \neq l$ are zero, and that only the coefficients of the diagonal terms can be nonzero, i.e., $B_{jl} = \delta_{jl} B_j$ and $D_{jl} = \delta_{jl} D_j$.

We will proceed from Eq. (II.4). After substituting in the solutions $\alpha_k(\omega_j)$ from (II.8) corresponding to the root ω_j^2 and dropping the constant b we find

$$\sum_{k=1}^{N} \left(\frac{A_{ik}}{\sqrt{m_i m_k}} - \delta_{ik}\omega_j^2 \right) \Delta_k(\omega_j^2) = 0, \quad i = 1, 2, \ldots, N.$$

(II.17)

Similarly for the other roots $\omega_l \neq \omega_j$

$$\sum_{k=1}^{N} \left(\frac{A_{ik}}{\sqrt{m_i m_k}} - \delta_{ik}\omega_l^2 \right) \Delta_k(\omega_l^2) = 0, \quad i = 1, 2, \ldots, N.$$

(II.18)

Multiplying the first equation of the system (II.17) (i = 1) by $\Delta_1(\omega_l^2)$, the second (i = 2) by $\Delta_2(\omega_l^2)$, etc., and adding all of the equations together

$$\sum_{i,k=1}^{N} \left(\frac{A_{ik}}{\sqrt{m_i m_k}} - \delta_{ik}\omega_j^2 \right) \Delta_k(\omega_j^2)\, \Delta_i(\omega_l^2) = 0.$$

(II.19)

We proceed similarly with the system of Eqs. (II.18)

$$\sum_{i,k=1}^{N} \left(\frac{A_{ik}}{\sqrt{m_i m_k}} - \delta_{ik}\omega_l^2 \right) \Delta_i(\omega_j^2)\, \Delta_k(\omega_l^2) = 0.$$

(II.20)

Since $A_{ik} = A_{ki}$ and the summation indices are arbitrary we can interchange the indices i and k in (II.19) and write this equation in the form

$$\sum_{i,k=1}^{N} \left(\frac{A_{ik}}{\sqrt{m_i m_k}} - \delta_{ik}\omega_j^2 \right) \Delta_i(\omega_j^2)\, \Delta_k(\omega_l^2) = 0.$$

(II.19a)

Subtracting (II.19a) from (II.20), we find

(II.21)

$$(\omega_j^2 - \omega_l^2) \sum_{i,k=1}^{N} \delta_{ik}\Delta_i(\omega_j^2)\, \Delta_k(\omega_l^2) = 0.$$

For $j \neq l$ and $\omega_j^2 \neq \omega_l^2$ it follows that

$$\sum_{k=1}^{N} \delta_{lk} \Delta_i(\omega_j^2) \Delta_k(\omega_l^2) = \sum_k \Delta_k(\omega_j^2) \Delta_k(\omega_l^2) = D_{jl} = 0, \quad \text{if} \quad j \neq l, \tag{II.22}$$

and we can write $D_{jl} = \delta_{jl} D_j$.

Taking this relation into account it follows from (II.20) that (the index l is not summed in this equation and ω_l^2 can be taken out from under the summation sign)

$$\sum_{i,k=1}^{N} \frac{A_{ik}}{\sqrt{m_i m_k}} \Delta_k(\omega_i^2) \Delta_i(\omega_j^2) \equiv B_{jl} = 0, \quad \text{if} \quad l \neq j. \tag{II.23}$$

We have now proved that only diagonal terms appear in the sums (II.15) and (II.16) over the normal coordinates, and the energy E of the system can be written in the form*

$$E = T + V = \frac{1}{2} \sum_j (B_j q_j^2 + D_j \dot{q}_j^2)$$
$$= \sum_j E_j (\text{harm. osc.}), \quad j = 1, 2, \ldots, N. \tag{II.24}$$

Thus the energy of a system undergoing small vibrations is a sum of harmonic oscillator energies E_j. The absence of terms which simultaneously depend on the coordinates of more than one oscillator shows that there is no interaction between the oscillators.[†]

We note that the introduction of normal coordinates in the description of small vibrations leads to a substantially simpler problem — the motion of independent harmonic oscillators. The j-th oscillator is a normal mode with frequency ω_j. Actually, from the expression for the energy of the j-th oscillator

$$E_j = \frac{D_j}{2} \left(\dot{q}_j^2 + \frac{B_j}{D_j} q_j^2 \right) \tag{II.25}$$

---

* The freedom in choosing the constant factors also allows us to find $\Delta_k(\omega_j^2)$ such that all of the coefficients $D_j = 1$, i.e., $\sum_k \Delta_k(\omega_j^2) \Delta_k(\omega_l^2) = D_{jl} = \delta_{jl}$. Usually the $\Delta_k(\omega_l^2)$ are chosen so that this relation is satisfied (see section II.5, Eq. (II.45)).

† Such interaction appears if we consider anharmonicity of the vibrations, i.e., if the expression for the potential energy of the crystal (II.1) also contains higher terms in the expansion in small displacements (see section II.4).

(by differentiating with respect to time) we can obtain the equation of motion for a harmonic oscillator

$$\ddot{q}_l^2 + \frac{B_l}{D_l} q = \ddot{q}_l + \omega_l^2 q_l = 0, \qquad (\text{II.25a})$$

in which the role of the square of the frequency is played by the ratio

$$\frac{B_l}{D_l} = \frac{\displaystyle\sum_{i,k} \frac{A_{ik}}{\sqrt{m_i m_k}} \Delta_k(\omega_l^2) \Delta_l(\omega_l^2)}{\displaystyle\sum_k \Delta_k(\omega_l^2) \Delta_k(\omega_l^2)} = \omega_l^2. \qquad (\text{II.26})$$

Here we used the definitions (II.15a) and (II.16a) and Eq. (II.19) [or (II.20)] with $l = j$.

$$\sum_{i,k=1}^{N} \frac{A_{ik}}{\sqrt{m_i m_k}} \Delta_k(\omega_l^2) \Delta_i(\omega_l^2) = \omega_l^2 \sum_{i,k=1}^{N} \delta_{ik} \Delta_k(\omega_l^2) \Delta_i(\omega_l^2) = \omega_l^2 \sum_k \Delta_k(\omega_l^2) \Delta_k(\omega_l^2).$$

$$(\text{II.27})$$

We note that relation (II.26) does not allow one to calculate ω_j^2 without solving the characteristic equation — one has to know the quantities $\Delta_k(\omega_j^2)$ which in turn can only be found by finding the eigenfrequencies ω_j^2. In order to find these one must solve the characteristic equation, which is already difficult for comparatively small molecules.

II.3   Repeated Roots of the Characteristic Equation

If the characteristic equation (II.5) has repeated roots then the above method for finding the solution (II.11) is not suitable and one must use other methods. These methods are well known in linear algebra [195] and in principle no difficulties appear. The existence and basic features of the normal-mode description of the vibrations also persists.

For repeated roots of the N equations (II.4), fewer than N−1 equations will be independent. If, for example, there is one two-fold root then two of the equations (II.4) are consequences of the others and there is not one constant a_N but two (which we call a_N

and a_{N-1}). The normal coordinates can be introduced as usual but some of them are not uniquely determined — linear combinations of normal coordinates having the same frequency ω_j are also normal coordinates. For example, it can happen that only two roots of the characteristic equation are equal, while all the rest remain distinct. We let $\omega_1^2 = \omega_2^2$ (we can number the roots in any manner). Then all of the remaining normal modes j (j = 3, 4, ... , N) have the same properties as when there are no repeated roots of the characteristic equation — each of them has its own frequency which does not coincide with other frequencies, $\omega_j \neq \omega_j'$ if j $\neq$ j', and its form (i.e., the pattern of cartesian displacements which corresponds to the given normal mode) is determined uniquely. The forms of the normal modes (coordinates) q_1 and q_2 are not uniquely determined: the linear combination $q = aq_1 + bq_2$ where a and b are arbitrary constants can serve as a normal mode of frequency ω_1 just as well as the initial q_1 and q_2.

Identical frequencies do not occur at random in a system of normal modes — they are a consequence of the symmetry properties of the vibrating system. Group theory allows one to establish the multiplicity of the eigenfrequencies and the forms of the normal modes, proceeding from the symmetry of the molecule or crystal, without solving the characteristic equation. Group theory does not give numerical values of the frequencies and the forms of the normal modes having the same frequency are not uniquely determined.

II.4   Classical Pattern of Small Vibrations

The vibrational pattern of a crystal or molecule depends not only on the dynamical properties of the system (mass and force constants) but also on the initial conditions — on what motion is given to the system at the initial instant of time. After introducing normal coordinates the dynamical properties of the system are reflected in the set of eigenfrequencies ω_j^2 (j = 1, 2, ... , N) and in the set of coefficients $\Delta_k(\omega_j^2)$ (k, j = 1, ... , N) which relate the cartesian coordinates to the normal coordinates by (II.11). There are 2N integration constants which characterize the "initial impetus" — the amplitudes c_j and phases α_j of the j-th normal modes.

As previously noted, the initial conditions can be chosen so that the whole system vibrates in one of the normal modes, say the

l-th. Here we must set all of the amplitudes except the l-th to be zero in (II.11), i.e., we set $c_j = \delta_{jl} c_l$. The equation which describes the motion of one point along one of the cartesian axes takes the form*

$$x_k(t) = \Delta_k(\omega_l^2) m_k^{-1/2} c_l \cos(\omega_l t + \alpha_l), \ k = 1, 2, \ldots, N. \qquad (II.28)$$

In this case all of the material points in the system (atoms and ions in the crystal) will undergo harmonic motion with a circular frequency ω_j. It can be seen from (II.28) that the displacement amplitude of a particular material point from its equilibrium position along a certain cartesian axis is proportional to the amplitude c_l of the normal mode (and the coefficient $\Delta_k(\omega_l^2)$). The amplitude c_l is the same for all k (i.e., for all material points and for all directions of the displacement from the equilibrium position). The quantity $\Delta_k(\omega_l^2)$, generally speaking, is different for different k. This shows, in particular, how the k-th material point participates in the l-th normal mode.

In order to excite only one (the l-th) normal mode, we can proceed as follows. We give all of the material points of the system displacements $A m_k^{-1/2} \Delta_k(\omega_l^2)$, where A is a factor common to all k, and the velocities are zero. We fix the material points at positions displaced according to the function $A m_k^{-1/2} \Delta_k(\omega_l^2)$. Then, at an instant t_0 which we choose as the origin $t_0 = 0$, we free the system from external forces and it begins to vibrate in the l-th normal mode according to Eq. (II.28), with a phase $\alpha_l = 0$, an amplitude $c_l = A$, and a frequency ω_l. If the potential energy of the system corresponds exactly to Eq. (II.1) this vibration will persist for an infinite period of time. The remaining normal modes would remain unexcited.

In a real system there is always some factor which brings about a coupling between the normal modes, so that the energy initially imparted to one of the normal modes ultimately is transferred to other normal modes. First we recall that the expression for the vibrational potential energy (II.1) is approximate: the higher terms in the power series expansion have been dropped. These terms (called the anharmonic terms in the potential energy) alter the pattern of vibrations described above.

---

* Remember that the index k is used to characterize both the number of the particle and the index of the cartesian axis.

Usually the effect of anharmonicity can be regarded as a small perturbation. Then the vibrations of the system can as usual be described on the basis of normal modes, but they are now not completely independent: there is a slow transfer of energy (slow relative to the vibration periods). Note that anharmonicity of the kind considered does not lead to damping of the vibrations of the system as a whole. Actually in considering the higher terms in the expansion of the potential energy, the vibrating system remains conceptually closed, and the vibrational energy is conserved. However, from the point of view of a single normal mode, damping does occur: if only the l-th normal mode is excited at t = 0 then in the course of time energy will be transferred to other normal modes and the amplitude of the l-th vibration will be damped until all of the normal modes attain more-or-less identical energies. If the system contains many particles and thus has a large number of normal modes, the amplitude of the l-th vibration decreases by a large factor during the relaxation.*

We cannot fully isolate the system from external influences, which also bring about damping of the vibrations. This refers to the effect of the surrounding medium as a heat reservoir, and the interaction with the electromagnetic field if the vibrating particles are charged, which leads to radiation of energy.

We must also keep in mind that in real atomic systems, in addition to the vibrational motion of the framework of the molecule (crystal) there is also electronic motion.

This interaction should be described on the basis of quantum theory. We stress that in real systems which undergo small vibrations, there are always effects which lead to coupling between normal modes. In large systems it is also possible to have relaxation processes of a different kind which occur without transfer of energy between normal modes.

The initial phase α_l of the l-th normal mode considered separately in Eq. (II.28) above is of no significance in the physical picture of the vibrations — by shifting the time origin we can make $\alpha_l = 0$. However, for the overall vibrational pattern of a system in which more than one normal mode is excited, the

---

* The decay law for the amplitude $c_l(t)$ is complicated in general. In particular, during relaxation a situation can arise which represents beats between weakly coupled harmonic oscillations.

initial phases play a fundamental role. Actually, the resulting
vibrational pattern can be regarded as interference of normal
modes — interference effects depend fundamentally on the phases
of the vibrations in addition to depending on the amplitudes.

We consider as an example the following initial arrangement
of the vibrations. Using external forces we set things up so that
at t = 0, in a system having a large number n of particles, the
material point number k = k' is displaced from its equilibrium
position by x_0, i.e., $x_{k'}(0) = x_0$ and $\dot{x}_{k'}(0) = 0$. The other points of
the system remain at their equilibrium positions. In normal co-
ordinates such an initial condition is described as follows according to (II.13) and (II.14)

$$m_k^{1/2} x_0 = \sum_{j=1}^{N} \Delta_{k'}(\omega_j^2) q_j(0) = \sum_{j=1}^{N} \Delta_{k'}(\omega_j^2) c_j \cos \alpha_j,$$

$$0 = \sum_{j=1}^{N} \Delta_k(\omega_j^2) c_j \cos \alpha_j \text{ with } k \neq k',$$ (II.29)

$$0 = m_k^{1/2} \dot{x}_k(0) = - \sum_{j=1}^{N} c_j \omega_j \Delta_k(\omega_j^2) \sin \alpha_j \quad \text{for all } k.$$

This system of equations determines both the amplitudes c_j
and the phases α_j of the normal vibrations. They are such that the
total effect of the normal vibrations on all of the material points
other than the k'-th is zero. This situation can occur only for a
particular choice of the initial phases α_j. Here in general all of
the $c_j \neq 0$, i.e., in the initial state under consideration none of the normal modes are in their equilibrium positions.*

If we now turn off the external force and let the system go,
it begins to vibrate. The amplitudes c_j remain as before but the
phases of the normal modes vary as $\omega_j t + \alpha_j$, which because of
differences in the frequencies ω_j is different for different normal
modes. Therefore the phase relations, which fully compensate the
action of the normal modes at t = 0 everywhere except at the k'-th
material point, vary with time so that an ever greater number of
points is involved in the motion. Ultimately all of the material
points will be involved more—or—less equally in the general motion

---

* Group theory allows one to find which $c_j = 0$ from symmetry considerations and to
  establish quantitative relations among the remaining c_j.

of the system. The vibrational energy, which at t = 0 was concentrated as potential energy near the k'-th particle, becomes distributed more—or—less uniformly as potential and kinetic energy over the whole system of n particles in the course of time. The vibrational energy of the k'-th particle thus relaxes. On the other hand, there is no redistribution of energy among the normal modes since each of them undergoes undamped simple harmonic motion with an energy determined by the time–independent amplitude c_l and the frequency ω_l. The natural energy process for an individual normal mode consists of the periodic transformation of potential energy into kinetic energy and back again in correspondence with the vibration law. The reason the vibrational pattern changes with time for the system as a whole is that the phase relations between the normal modes change.

More detailed investigations of this problem, carried out because it is important for the theory of lattice vibrations, show that there is a spreading of the wave packet of normal modes. The relaxation rate of the initial displacement x_0 is greater the greater the range of normal mode frequencies ω_j which have nonzero amplitudes c_j. For a system with a large number of particles (for example, a crystal) the frequencies ω_j are quite dense and we can consider the amplitudes c_j of the normal modes in the packet to be a function $c(\omega)$ of the continuous variable ω. The function $c(\omega)$ often turns out to be a function* with one maximum. Then the characteristic relaxation rate of the displacement x_0, or what is the same thing, the spreading rate of the wave packet of normal modes, corresponds to the half-width of the function $c(\omega)$. The greater the half-width the more rapidly relaxation occurs. Qualitatively this result is quite clear. The greater the differences among the frequencies ω_j of the normal modes displaced from equilibrium by a given initial displacement x_0, the faster (in comparison to the normal mode period near the maximum of $c(\omega)$) the phase relations between them are destroyed.

The relaxation process under consideration for the displacement x_0 occurs under conditions where the system undergoes harmonic vibrations.

---

* For a crystal this occurs, strictly speaking, for each of the allowed vibrational forbidden bands separately.

The packet will be narrower and the relaxation of an initially localized vibrational state will be slower if we displace more than one particle in a particular way. For example, for a localized vibration about an impurity in a crystal one can choose displacements which reproduce the localized vibration of the impurity and some of its neighboring particles. In the limit we can give cartesian displacements such that we excite one normal mode exactly. But to do this we must displace all of the particles in the system from their equilibrium positions (even for exact reproduction of a localized vibration!). In this limiting case the wave packet becomes infinitesimally wide — it is composed of a single normal mode (or normal modes having the same frequency). The vibrational pattern will correspond to excitation of one normal mode, and in the harmonic approximation there is no relaxation.

Anharmonicity of the vibrations (or some of the other causes mentioned above) leads to additional relaxation of localized vibrational excitation due to energy transfer between normal modes. If the wave packet of normal modes is wide in comparison to the anharmonic width $\Delta\omega_j$ of a single normal mode in the packet* then the localized excitation relaxes by spreading of the wave packet (harmonic relaxation mechanism). The initially localized vibrational excitation spreads rapidly over the crystal. The redistribution of energy between normal modes arising from anharmonicity occurs in parallel but more gradually. The interaction between normal modes "succeeds" in affecting the relaxation process only when the localized vibration has already spread out.

If the wave packet of normal modes is narrower than $\Delta\omega_j$ then relaxation arising from anharmonicity predominates.

In conclusion, we clarify our remarks about the validity of the transition from a discrete set of eigenfrequencies ω_j to a continuous spectrum ω. Strictly speaking, this transition changes the nature of the relaxation process in principle, and if there is no external anharmonic relaxation[†] the transition to a continuous frequency spectrum is inadmissable. Indeed, harmonic relaxation is

---

* The anharmonic width $\Delta\omega_j$ can be estimated from the relaxation $\Delta\omega_j \approx \tau_j^{-1}$, where τ_j is the characteristic time associated with anharmonic relaxation of the j-th normal mode.

† By external anharmonicity we mean all of the causes considered above which lead the system eventually to thermal equilibrium.

in essence a beat process. With a finite number n of particles in the system there is always a general period T for all of the normal modes and therefore in fact the localized excitation relaxes and undergoes a certain periodic beat process with a period T, after which the wave packet again narrows. With large n the period T becomes enormous but in principle undamped beats remain. With a continuous frequency spectrum, which corresponds to $n \to \infty^*$, T becomes infinite and as a result, instead of the beats between the harmonic vibrations, there is true relaxation of the localized vibrational excitation.

We obtain the physical picture of relaxation described above by considering external anharmonicity, which is also justification (from physical considerations) of the transition to a continuous frequency spectrum. Actually, if a system contains enough particles that the beat period T is many times greater than the external anharmonic relaxation time, the spreading at the start of the beat process of the wave packet never goes backward and the relaxation is irreversible.

The reader will find a more detailed treatment of harmonic relaxation processes, both in classical and quantum-mechanical form, in [61, 196]. Note that such relaxation of the vibrational energy in a quantum treatment is, in essence, a particular case of isoenergetic decay of a quasistationary quantum state [197, 198].

Our description of the basic features of the relaxation of localized vibrational excitation persists for a quantum treatment of the vibrations also. In particular, the time dependence of the vibrational energy in a localized region ("outflow law for the excess energy") is the same for both classical and quantum-mechanical theories [61, 196].[†]

---

* A continuous spectrum corresponds, strictly speaking, to vibrations in an infinite continuous medium.

† In a quantum picture one can no longer speak of exact initial values of the coordinates and momenta which are simultaneously defined. One also needs to clarify the discussion of the magnitudes of the potential and kinetic energies and their mutual transformation. In the quantum picture it is appropriate to speak of the average values of these quantities. Furthermore, Purga [196] deals with the excitation of a center by a pulsed, strongly nonmonochromatic photon packet. During excitation by strongly monochromatic light, vibrational relaxation does not occur since the vibronic transitions already have excited a very broad region of the crystal so that the thermal equilibrium of the impurity center is hardly disturbed [169].

II.5   Normal Coordinates and the Quantum-Mechanical Problem of Small Vibrations [48]

To acquaint the reader with the approaches used in present-day quantum-mechanical theories of solids, we approach the problem from another point of view — as separation of variables in the Schrödinger equation by simultaneous reduction of the potential and kinetic energies to diagonal form.

We proceed from the expression for the potential energy (II.1a) (dropping the unimportant constant term)

$$V = \frac{1}{2} \sum_{i,k=1}^{N} A_{ik} x_i x_k.$$ (II.30)

It is convenient to introduce the reduced displacements using the equation

$$u_i \equiv m_i^{1/2} x_i.$$ (II.31)

Then the potential energy can be written in the form

$$V = \frac{1}{2} \sum_{i,k=1}^{N} D_{ik} u_i u_k,$$

where the quantities

$$D_{ik} \equiv (m_i m_k)^{-1/2} A_{ik}$$ (II.32)

form an $N \times N$ matrix $\hat{D}$ called the dynamical matrix.

Our aim is to obtain the Schrödinger equation which describes the vibrations. We choose a coordinate representation in which the coordinates are multiplicative operators. We have for the potential energy operator

$$\hat{V} = \frac{1}{2} \sum_{i,k=1}^{N} D_{ik} \hat{u}_i \hat{u}_k,$$ (II.33)

where $\hat{u}_i$ are multiplicative operators. We have for the kinetic energy

$$\hat{T} = \frac{1}{2} \sum_{i=1}^{N} m_i \hat{\dot{x}}_i^2 = \frac{1}{2} \sum_{i=1}^{N} \dot{u}_i. \tag{II.34}$$

Since x_i and u_i are rectilinear coordinates, the reduced momenta u_i in the expression for the kinetic energy can be replaced by partial differentiation with respect to the reduced coordinates to which the given momenta are canonically conjugate. We find that

$$\hat{T} = -\frac{1}{2} \sum \hat{p}_i^2,$$

where   (II.35)

$$\hat{p}_i \equiv -ih \frac{\partial}{\partial u_i}$$

is the momentum operator canonically conjugate to the coordinate u_i. The hamiltonian operator $\hat{H}$ which enters into the Schrödinger equation $\hat{H}\psi = E\psi$ takes the form

$$\hat{H} = \hat{T} + \hat{V} = \frac{1}{2} \left(-h^2 \sum_{i=1}^{N} \frac{\partial^2}{\partial u_i^2} + \sum_{i,k=1}^{N} D_{ik} u_i u_k \right). \tag{II.36}$$

Note that because we have introduced the reduced displacements and the dynamical matrix, the masses of the nuclei do not enter explicitly in the expressions given above. The coordinates u_i are dynamically related to each other since the potential energy operator contains cross terms $u_i u_k$. Therefore the motion of the mutually coupled coordinates is complicated. We do not directly solve the Schrödinger equation with the hamiltonian (II.36) expressed in the coordinates u_i. In order to get rid of the cross terms in (II.36) and reduce the Schrödinger equation to a form in which the variables are separated we introduce the new set of coordinates q_s (s = 1, 2, ... , N) which are related by a linear transformation to the u_i

$$q_s = \sum_i e_{is} u_i. \tag{II.37}$$

In matrix notation

$$q = u\hat{e}, \tag{II.37a}$$

where q and u are N-dimensional vectors and $\hat{e}$ is an N×N matrix with elements e_{is}. The inverse transformation can be written as

$$u = q\hat{e}^{-1}, \qquad \text{(II.37b)}$$

where

$$\hat{e}^{-1}\hat{e} = 1, \qquad \text{(II.37c)}$$

or in the form of a relation between the vector components

$$u_i = \sum_s e_{si}^{-1} q_s. \qquad \text{(II.37d)}$$

We see that the problem of choosing these transformation coefficients e_{is} so that the hamiltonian operator in the new coordinates does not contain terms of the type $q_i q_k$ (i ≠ k) is a mathematical problem fully equivalent to the classical mechanical problem of finding the eigenfrequencies and normal coordinates of a system undergoing small vibrations considered above.*

We require that there be no nondiagonal terms in the potential energy operator in the new coordinates. Then we show that the resulting requirements on the properties of the transformation coefficients guarantee that the diagonal form of the kinetic energy operator is preserved.

In the new coordinates the potential energy takes the form

$$\sum_{i,k} D_{ik} u_i u_k = \sum_{i,k} D_{ik} \sum_{s,l} e_{li}^{-1} q_l e_{sk}^{-1} q_s.$$

The transformation coefficients in (II.37d) should be chosen so that the relation

$$\sum_{s,i,k,l} D_{ik} e_{li}^{-1} e_{sk}^{-1} q_l q_s = \sum_s \sum_{i,k,l} D_{ik} e_{li}^{-1} e_{sk}^{-1} q_l q_s = \sum_s \omega_s^2 q_s^2,$$

is satisfied, or

$$\sum_{i,k,l} D_{ik} e_{li}^{-1} e_{sk}^{-1} q_l q_s = \omega_s^2 q_s^2. \qquad \text{(II.39a)}$$

---

* In the final analysis we find that normal coordinates can be introduced with the same success as in the classical expression for the energy of a vibrating system when we turn to the quantum-mechanical expression for the energy operator.

Here ω_s^2 are constants having the dimensions of squares of frequencies, which are to be determined.

Equations (II.39a) will be satisfied if we can satisfy the system of equations

$$\sum_{i,k} D_{ik} e_{li}^{-1} e_{sk}^{-1} = F_s \delta_{sl}, \quad s, l = 1, 2, \ldots, N, \tag{II.39b}$$

where F_s are certain constants.

These equations will be satisfied if two sets of conditions

$$\sum_k D_{ik} e_{sk}^{-1} = G_s e_{si}^{-1}, \tag{II.40a}$$

$$\sum_i e_{si}^{-1} e_{li}^{-1} = H_s \delta_{sl}, \tag{II.40b}$$

are fulfilled, where G_s, H_s, and $F_s = G_s H_s$ are constants. The simplest thing is to set $G_s = \omega_s^2$ and $H_s = 1$ so that we have

$$\sum_k D_{ik} e_{sk}^{-1} = \omega_s^2 e_{si}^{-1}, \quad i = 1, 2, \ldots, N, \tag{II.41a}$$

$$\sum_i e_{si}^{-1} e_{li}^{-1} = \delta_{sl}, \quad s, l = 1, 2, \ldots, N. \tag{II.41b}$$

Using Eqs. (II.37c) it is not difficult to establish that if (II.41b) is fulfilled, we have the simple relation

$$e_{is}^{-1} = e_{si}, \tag{II.42}$$

between the elements of the matrices $\hat{e}$ and $\hat{e}^{-1}$, i.e., the matrix for the inverse transformation $\hat{e}^{-1}$ is the transpose of $\hat{e}$. We can rewrite (II.41a,b) as equations for determining the elements of the direct transformation matrix (II.37)

$$\sum_k D_{ik} e_{ks} = \omega_s^2 e_{is}, \quad i = 1, 2, \ldots, N, \tag{II.43a}$$

$$\sum_i e_{is} e_{il} = \delta_{sl}, \quad s, l = 1, 2, \ldots, N. \tag{II.43b}$$

Condition (II.43a) is a system of N linear homogeneous differential equations, where the unknowns are the transformation coefficients e_{is} (i = 1, 2, ... , N) and the ω_s^2 are to be chosen so that the system of equations is soluble. It is known from algebra that the condition for solubility is the vanishing of the determinant of coefficents for the system (II.43a)

$$| D_{ik} - \omega_s^2 \delta_{ik} | = 0. \tag{II.44}$$

The equation obtained is the characteristic equation, which coincides with Eq. (II.5). It is an equation of N-th degree in ω_s^2, whose coefficients are polynomials in the elements of the dynamical matrix. The index s of the constants ω_s^2 (s = 1, 2, ... , N) which we had to introduce in (II.38) to describe the diagonal potential energy, in general numbers the solutions of the characteristic equation (II.44). The solutions are actually N in number, i.e., exactly the number we needed. For each of the solutions ω_s^2 of Eq. (II.43a) there is a set of quantities e_{is} (i = 1, 2, ... , N).

The quantities e_{is} are not uniquely defined. Multiplication of all the e_{is} by a common factor gives a new set of e_{is} which is also a solution of Eqs. (II.43a). Additional arbitrariness in the definition of e_{is} appears if some of the roots are equal (degenerate case). We can, however, show that a set e_{is} compatible with (II.43a) can always be chosen so that relation (II.43b) is satisfied, along with another analogous relation

$$\sum_i e_{is} e_{il} = \delta_{sl}; \ \sum_k e_{sk} e_{lk} = \delta_{sl}. \tag{II.45}$$

Actually Eq. (II.43a) can be regarded as an equation for the eigenfrequencies and eigenvectors e_{is} of the dynamical matrix D_{ik}; Eq. (II.45) is the orthogonality and normalization condition* for the matrix $\hat{e}$ formed from the eigenvectors e_{is} for all s. It is known from linear algebra that the orthogonalization and normalization procedure is always possible for a system of eigenvectors of a matrix having the properties of the dynamical matrix (II.30) and (II.32) are symmetric positive definite quadratic forms). From a comparison with the equations of sections II.1 and II.2 it is evident that there is a connection between e_{is} and $\Delta_i(\omega_s^2)$; in general they agree up to the arbitrariness in the choice of the systems (II.4) and (II.43a), i.e., in the absence of degenerate eigenfrequencies ω_s with a precision up to the normalization, and in the presence of degeneracy with a precision up to the choice of the linear combinations of solutions corresponding to degenerate eigenfrequencies. The condition (II.45) is equivalent to the choice $D_j = 1$ in Eq. (II.22). We can set $e_{is} = \Delta_i(\omega_s^2)$ [if after finding $\Delta_i(\omega_s^2)$ we also carry out a normalization procedure and orthogonalize the solutions for degenerate frequencies.

---

* If e_{is} is complex, (II.45) expresses the unitarity of the transformation.

Thus there is always a set e_{is} which simultaneously satisfies Eqs. (II.43a) and (II.45). The quantities e_{is} form a matrix $\hat{e}$. Equation (II.45) implies that this matrix is nonsingular and orthogonal. This choice of the e_{is} (regardless of whether it is the only possible one or one of a set of possible choices) can be used to determine the transformation (II.37) to the normal coordinates q_s which reduce the potential energy operator in (II.36) to diagonal form. It remains to be shown that the kinetic energy operator does not contain any cross terms when we transform to the coordinates q_s.

To express the kinetic energy operator in normal coordinates we consider its action on an arbitrary function

$$\varphi(q_1(u_1, u_2, \ldots, u_N);\ q_2(u_1, u_2, \ldots, u_N);\ \ldots;\ q_N(u_1, u_2, \ldots, u_N)),$$

in which the normal coordinates q_j are related by (II.37) to the reduced displacements u_i. According to the chain rule for differentiation we have

$$\hat{T}\varphi = -\frac{h^2}{2}\sum_i \frac{\partial^2\varphi}{\partial u_i^2} = -\frac{h^2}{2}\sum_i\sum_s \frac{\partial}{\partial q_s}\left(\sum_l \frac{\partial\varphi}{\partial q_l}\frac{\partial q_l}{\partial u_i}\right)\frac{\partial q_s}{\partial u_i} =$$

$$-\frac{h^2}{2}\sum_{l,s}\frac{\partial}{\partial q_s}\left(\sum_l \frac{\partial\varphi}{\partial q_l}e_{il}\right)e_{is} = -\frac{h^2}{2}\sum_{s,l}\frac{\partial^2\varphi}{\partial q_s\partial q_l}\delta_{sl} = -\frac{h^2}{2}\sum_s \frac{\partial^2\varphi}{\partial q_s^2}.$$

Here we use the fact that $\partial q_l /\partial u_i = e_{il}$ (II.37), change the order of the summation, and use Eq. (II.45).

Thus in the normal coordinates q the hamiltonian operator actually is a sum of terms each of which depends on only one normal coordinate

$$\hat{H} = \frac{1}{2}\sum_s\left(-h^2\frac{\partial^2}{\partial q_s^2} + \omega_s^2 q_s^2\right).$$

In the Schrödinger equation

$$\hat{H}\Phi(q_1, q_2, \ldots, q_N) = \frac{1}{2}\sum_s\left(-h^2\frac{\partial^2\Phi}{\partial q_s^2} + \omega_s^2 q_s^2\Phi\right) = E\Phi \qquad \text{(II.46)}$$

we separate variables as usual. Setting

$$\Phi(q_1, q_2, \ldots, q_N) = \varphi_1(q_1)\varphi_2(q_2) \cdots \varphi_N(q_N) = \prod_{s=1}^{N}\varphi_s(q_s), \qquad \text{(II.47)}$$

we find that (II.46) decomposes into N equations

$$\frac{1}{2}\left(-\hbar^2\frac{\partial^2}{\partial q_s^2}+\omega_s^2 q_s^2\right)\varphi_s(q_s)=\varepsilon_s\varphi_s(q_s),$$

(II.48)

where the conditions imposed on the separation parameters give the relation between the energy eigenvalues

$$E=\sum_{s=1}^{N}\varepsilon_s.$$

(II.49)

Each of Eqs. (II.48) is a Schrödinger equation for a simple harmonic oscillator with a circular frequency ω_s and unit mass. The solution of this equation is well known (see, for example, [70, 104]). The energy eigenvalues or energy levels of the harmonic oscillators assume the values

$$\varepsilon_s=\varepsilon_s(i_s)=\left(i_s+\frac{1}{2}\right)\hbar\omega_s,$$

(II.50)

where the quantum number i_s for the s-th oscillator can assume integer values beginning with zero, i.e., $i_s = 0, 1, 2, \ldots$.

The wave function of the s-th oscillator corresponding to the i_s-th energy level is denoted by

$$\varphi_{i_s}(q_s), \quad i_s=0, 1, 2, \ldots$$

(II.51)

It is, as is well known, a Hermite polynomial of the i-th degree multiplied by a gaussian weight function

$$\varphi_{i_s}(q_s)=(\bar{q}_s)^{-1/2}e^{-\xi_s^2/2}H_{i_s}(\xi_s).$$

(II.52)

Here we have introduced the dimensionless quantity $\xi_s = q_s(\bar{q}_s)^{-1}$ where $q_s \equiv (\hbar/\omega_s)^{1/2}$. The normalized Hermite polynomial H_{i_s} is defined by

$$H_{i_s}(\xi_s)=\frac{(-1)^{i_s}}{\sqrt{2^{i_s}i_s!\sqrt{\pi}}}e^{\xi_s^2}\frac{d^{i_s}}{d\xi_s^{i_s}}e^{-\xi_s^2}.$$

The quantum-mechanical stationary vibrational states of a system of material points which undergo small vibrations are given by the solutions of the full Schrödinger equation (II.46). According to Eqs. (II.49) and (II.50) the energy eigenvalues (energy levels) of the whole system are characterized by the set of quantum numbers $i_1, i_2, \ldots i_s, \ldots , \ldots , i_N$).

Thus we have for the total energy of the system

$$E(i) = E(i_1, i_2, \ldots, i_s, \ldots, i_N) = \sum_{s=1}^{N} \varepsilon_s(i_s) = \sum_{s=1}^{N} \left(i_s + \frac{1}{2} \right) \hbar \omega_s. \quad \text{(II.53)}$$

The wave function $\Phi_i(Q)$ which describes the system in the i-th stationary state (here i denotes the set of quantum numbers and Q is the set of normal coordinates) is a product of the corresponding wave functions for the individual oscillators (II.51)

$$\Phi_i(Q) = \Phi_{i_1,i_2,\ldots,i_N}(q_1, q_2, \ldots, q_N) = \varphi_{i_1}(q_1)\varphi_{i_2}(q_2) \ldots \varphi_{i_N}(q_N) = \prod_s \varphi_{i_s}(q_s). \quad \text{(II.54)}$$

II.6   The Effects of a Localized Change in the Dynamics on the Normal Modes

We will give a short discussion of the basic relations between normal coordinates and the cartesian displacements in an impurity center as outlined in Table 1.

We turn to the orthogonality (unitarity) conditions (II.41b), (II.43b), and (II.45) of the transformations (II.37) and (II.37d). We first consider the band vibrations.* They involve all of the crystal more-or-less identically, i.e., the cartesian displacements of the atoms, and according to (II.37d) and (II.42) the coefficients $e_{si}^{-1} = e_{is}$ too are of the same order of magnitude throughout the crystal. From Eq. (II.43b) with s = l, whose left-hand side consists of N terms of the same order of magnitude for band vibrations, it follows that

$$e_{is}^2 \sim N^{-1} \quad \text{(II.55a)}$$

and

$$e_{is} \sim N^{-1/2}. \quad \text{(II.55b)}$$

It is evident from Eq. (II.43b) with arbitrary s (s ≠ l) that the coefficients e_{is} (with a given i and different s = 1, 2, ... , N) have different signs or, of they are complex, different phases. Therefore it is not surprising that the sum (II.37d), which contains N terms each of which is of the order of $N^{-1/2}$, gives finite shifts

---

* We will discuss the vibrations in a way which does not require periodicity of the system.

u_i in the cartesian displacements. With periodic symmetry and a suitable choice of boundary conditions it is not hard to see that the phases of the coefficients e_{is} actually ensure finiteness of the u_i. In the absence of periodicity we might, to be explicit, assume that the sum (II.37d) is the path length L traversed after N random steps on a line or, if the coefficients are complex, in a plane. As is well known, in both cases L is of the order of $N^{1/2}$.

The order of magnitude of the coefficients of the bilinear terms in the normal coordinates is easily established by using the transformation (II.37d) twice, taking (II.55) into account.

$$e_{li}^{-1} e_{sk}^{-1} = e_{il} e_{ks} \sim N^{-1}. \tag{II.56}$$

In a localized normal mode only some finite number of atoms in the vicinity of the impurity participate, since the cartesian displacements decrease very rapidly (exponentially) with distance from the impurity. Therefore in the sum (II.43b) with $s = l$ we assume that only a finite number r of the terms e_{is}^2 are nonzero, each of which therefore has a finite value. The contribution of the remaining $N-r$ displacements can be neglected. Thus for a localized vibration e_{is} is finite if the index i corresponds to the cartesian displacement of an atom located in the "active region of the localized vibration;" for the indices i outside the active region the e_{is} are negligible.

Exercise 40. Show that the displacements u_i, which are calculated from (II.37d) for the normal modes of an ideal periodic crystal taking the condition $e_{is} \sim N^{-1/2}$ into account, are finite. Assume periodic boundary conditions.

Exercise 41. Using the transformation (II.37) and Eq. (II.55) show that finite shifts in the cartesian equilibrium positions of a finite number of atoms lead to a shift in the equilibrium position of a normal mode which is of the order of $N^{-1/2}$.

II.7 Suppression of the Band Vibrations in the Presence of Localized Vibrations

The "exclusion" of the band vibrations from the active region of a localized vibration is evident from the normalization condition (II.45); for the localized vibration the quantities e_{is}^2 are large for the atoms which participate in the localized vibration and the share for the band vibrations remains substantially less than in the absence of the localized vibration. Actually, taking (II.42) into account, Eq. (II.41b) can be written as

$$\sum_{\varkappa} e_{i\varkappa}^2 = 1 - e_{i\lambda}^2, \tag{II.57}$$

where i is the index of the atom and the cartesian displacement and $\varkappa$ and λ refer to the band and localized vibrations respectively.

In the presence of several localized vibrations we have

$$\sum_{\varkappa=L+1}^{N} e_{i\varkappa}^2 = 1 - \sum_{\lambda=1}^{L} e_{i\lambda}^2, \tag{II.58}$$

where L is the number of localized vibrations.

The stronger the participation of the i-th atom in localized vibrations the less it participates in the band vibrations.

The qualitative phenomenon of "exclusion" is easily interpreted. Actually if conditions are created for a localized vibration in the neighborhood of an impurity, i.e., for a vibration which cannot resonate with the band vibrations, then in this very region the conditions for propagation of the band modes are violated. The region around the impurity is capable of resonating with the localized frequency but cannot resonate with band vibrations.

II.8   Localized Lattice Dynamics in the Neighborhood of a Defect and Pseudolocalized Vibrations

In recent years considerable success has been achieved in understanding band vibrations in the vicinity of a point defect. It turns out that even in the absence of a localized vibration the distortion of the band vibrations about a defect can be quite considerable. The general nature of the phenomenon is easily understood physically; near an impurity with increasing interatomic force constant or decreasing impurity mass in comparison to the ion which it replaces, the vibration amplitudes are enhanced for the high-frequency vibrations (a "potential candidate for a localized vibration" begins to form) and decrease for the low-frequency vibrations. With increasing impurity mass or decreasing interatomic force constant the opposite occurs: the amplitudes of the low-frequency vibrations increase.

An investigation of the contributions from vibrations of various frequencies to the displacement of some individual s-th ion of the lattice in the presence of a defect (but in the absence of a localized vibration) leads to an important conclusion: The cartesian displacements $x_p(\omega)$ (and also $y_p(\omega)$ and $z_p(\omega)$) for an impurity ($p = 0$) and its neighbors ($p = 1, 2, \ldots, r$, where r is the number of neighbors whose vibrations are substantially perturbed by the introduction of the impurity) can have sharp maxima as a function of the normal mode frequency ω (Eq. (II.37c) or (II.14)) in a frequency range where there is a "smooth" distribution of $x_p(\omega)$ for the ideal crystal. The frequency range in which we would expect these additional maxima with a given kind of perturbation is determined by the rules given above. As the preturbation increases, at a certain point a localized vibration appears, and then its contribution to x_0 becomes substantial while the former maximum in $x_0(\omega)$ vanishes. If there are several vibrational bands $x_p(\omega)$ changes independently for different bands; there can be substantial maxima in $x_p(\omega)$ for one band, while others have already "given birth" to localized vibrations. If the growing perturbation due to the impurity does not shift a maximum of $x_p(\omega)$ toward a forbidden band in the phonon spectrum then in general no localized vibration occurs and the maximum becomes more pronounced. Such a situation arises, for example, in the acoustic vibration band for heavy impurities.

For a number of processes, including vibronic transitions, which occur in small-radius impurity centers with the participation of the surrounding particles of the lattice, linear combinations

$$X^{(r)} = \sum_{p=0}^{r} (c_{px}x_p + c_{py}y_p + c_{pz}z_p) \equiv \sum_p \sum_\gamma c_{p\gamma}u_{p\gamma}, \qquad \text{(II.59)}$$

of the displacements of a small number of atoms having a certain symmetry are important. Here the index γ denotes the components of the displacement vector. These combinations can be expressed in terms of the normal coordinates q_s of the lattice vibrations,

$$X^{(r)}(q_s) = X^{(r)}(\omega) = \sum_p \sum_\gamma \sum_s c_{p\gamma}e_{ps}^{(\gamma)}q_s, \qquad \text{(II.60)}$$

where the frequency dependence is determined by the fact that the normal coordinate q_s corresponds to a frequency ω_s. If the index p corresponds to an atom far from the impurity in an undistorted region of the crystal, then the function is hardly different from the corresponding function for the ideal crystal. These remarks also apply to linear combinations $X^{(r)}$ of atoms outside the defect region. However, for atoms near the impurity, as noted above, because the form of the vibration is distorted about the defect, $u_{p\gamma}(\omega)$ can have additional sharp maxima which do not occur in $u_{p\gamma}(\omega)$ for the undistorted region of the crystal. This means that of the matrix elements e_{is} for the transformation of the atomic displacements in the impurity center to normal coordinates, only those for which ω_s lies in a narrow frequency range $\Delta\omega$ are important.

It is to be understood that $X^{(r)}(\omega)$ can also have some such sharp maxima; depending on the problem under consideration and its symmetry (which establishes the particular set of coefficients $c_{p\gamma}$, and also the "interference conditions" for the contributions of the displacements $u_{p\gamma}$ of the various atoms to a given normal mode) the linear combinations exhibit some of the maxima of the functions $u_{p\gamma}(\omega)$ in somewhat changed form. The dependence of the results on the set $c_{p\gamma}$ means that different kinds of vibrations about the impurity play a role in different phenomena.

On the other hand, it is important to say something here about the frequency distribution of the normal modes. If it contains regions where the number of vibrations is large (many values of the index p correspond to about the same frequency) maxima can occur in $u_{p\gamma}(\omega)$ and $X^{(r)}(\omega)$. The picture becomes especially clear if we replace the quasidiscrete phonon spectrum by a frequency distribution function $\rho(\omega)$. Then we must consider the function $F(\omega) = X^{(r)}(\omega)\rho(\omega)$, the spectral density of the combination of ion displacements under consideration. The peaks in $F(\omega)$ which are important for a given process can come either from peaks in $X^{(r)}(\omega)$ or peaks in $\rho(\omega)$. If the perturbing effect of the impurity on the vibrations is strong the peaks in $F(\omega)$ may come from the peaks in $X^{(r)}(\omega)$. Evidently in this case the positions of the peaks can be substantially different for different electronic states of the impurity. If, however, the perturbing effect of the impurity is weak then the maxima of $F(\omega)$ will be close to the maxima in $\rho(\omega)$ and will depend only weakly on the electronic state of the impurity. Thus the motion of the impurity and its neighbors which is important for the physical phenomenon under consideration may be largely determined by the vibrations in a narrow part of the continuous phonon spectral function $F(\omega)$.

Narrow maxima in the function $F(\omega)$ should manifest themselves in various processes in many quite narrow vibrational bands, and these appear like localized vibrations. The vibrational states of a crystal which correspond to excited vibrational motion of the particles near an impurity with frequencies corresponding to the peaks in $X^{(r)}$ are called pseudolocalized vibrations.*

Various pseudolocalized vibrations having different symmetries can occur about a single defect center. They will also have different effective frequencies, which in general change by a finite amount with a change in the electronic state of the impurity.

The simplest case occurs in the Mössbauer effect: the composition of the actual vibrational wave packet is determined by the expansion coefficients of the displacements (in momentum space) of only the one nucleus which receives the gamma quantum recoil.

Moreover, in metals it can evidently be assumed to a good approximation that heavy impurities affect the lattice vibrations only with regard to their mass, while the interatomic force constants remain the same. In this case the pseudolocalized vibration problem can be completely solved, which has had a great impact on the development of the whole concept of pseudolocalized vibrations [31, 32]. Good agreement with experiment was later shown [199].

Pseudolocalized vibrations, in contrast to localized vibrations, are not normal modes of the crystal. They can be represented as wave packets of normal modes and consequently they are not stationary even in the harmonic approximation for the lattice vibrations. Anharmonic processes are needed for relaxation of a localized vibration, in which its quanta decay into two, three, or more phonons. Therefore the relative (for the same vibrational frequency) lifetime of a pseudolocalized vibration will be less than the lifetime of a localized vibration and it is to be expected that

---

* Sometimes the name pseudolocalized vibration is given to all narrow maxima of $F(\omega)$, i.e., the regions of dense frequencies in the function $\rho(\omega)$ are also called pseudolocalized vibrations. This is a matter of definition.

the localized vibrations will give appreciably narrower lines in the vibrational structure of a vibronic spectrum, especially at low temperatures.

To estimate the relative lifetimes we can take $\tau/T = h\omega/\Delta E = \omega/\Delta\omega$ where τ is the lifetime, T is the period, $\Delta\omega$ is the frequency half-width of the vibrational line in the spectrum, and ω is the distance between the lines in the series corresponding to the localized or pseudolocalized vibration under consideration.

Investigations of the relaxation of a displacement initially imparted to one of the neighbors of the impurity or the impurity itself, using the simplest models, show that it can occur substantially faster than in the ideal lattice. The relaxation time is greater the more pronounced the peak in the spectral density function of the displacement, i.e., the "closer" the pseudolocalized vibration is to a localized one. It can become of the order of several tens of vibration periods, while in the ideal lattice the characteristic relaxation time is of the order of several vibration periods [61, 196].*

In speaking of the relaxation of pseudolocalized vibrations we must not, ultimately, forget that we deal with a quasi-stationary quantum-mechanical state whose relaxation is influenced not only by the hamiltonians for the system and the perturbation, but by the initial conditions too (see [197, 198]). If the width of the initial vibrational wave packet substantially exceeds the widths of the vibrational levels arising from anharmonicity, nonadiabaticity, and other interactions which can be regarded as perturbations in a harmonic description of the vibrations, then the essential first relaxation stage of the wave packet† corresponds to just the inherent quasistationariness of the state, i.e., to the composition of the wave packet constructed of eigenfunctions of the harmonic vibrational hamiltonian. In other words, the relaxation process is governed by the method of excitation. In our case the vibrational relaxation in the impurity center depends on the form of excitation. In particular, in photoexcitation there should be a dependence of

----

* The related problem of energy propagation in a linear chain is treated in [200].

† The perturbations begin to play a role during further stages of relaxation, when the pseudolocalized vibration has in fact already decayed — the vibrational perturbation almost completely leaves the impurity center. Some estimates have been made in [196].

the relaxation on the spectral composition of the individual "photon" in the light flux. To excite the full vibrational packet which corresponds to a pseudolocalized vibration it is necessary to have "quite white" photons which have a spectral width exceeding the width of the corresponding vibrational band. The absorption of a "more-or-less monochromatic" photon leads to excitation of vibration in a somewhat larger region of the crystal so that vibrational relaxation should be regarded as occurring "from the time of the absorption event" (see [169]). As we know, at present there is no theory which allows one to reliably judge the degree of monochromaticity of individual photons from a real light source, or the corrections to the process of exciting the center coming from effects of the whole crystal acting as a heat bath. These problems are related to the interesting problem of quantum beats and to the fundamental problems of the quantum theory.

The detailed theory gives a more precise description of the vibrational situation in an impurity center than a simple set of localized and pseudolocalized vibrations. The original exact theory gives expressions like (II.59) and (II.60) which lead to the complete frequency distribution of the "interaction function" with the vibrations for any physical phenomenon. In the Mössbauer effect this is the distribution of the recoil energy among the vibrations, and for vibronic phototransitions it is the distribution of the Stokes loss among the vibrations. The concept of a pseudolocalized vibration is important and quite useful for a qualitative understanding of the physical situation, but ultimately it is an approximate, auxiliary idea.

Appendix III

Some Sum Formulas

The following formulas are useful in the theory of quasiline vibronic spectra:*

1. The formula for calculating the intensity of a quasiline taking localized vibrations into account is

$$\sum_m J^2_{m,m+n} q^{2m} = \frac{q^{-n}}{1-q^2} \exp\left(-\frac{1+q^2}{1-q^2}\frac{x_0^2}{2}\right) I_n\left(\frac{q}{1-q^2}x_0^2\right), \quad q^2 < 1. \quad \text{(III.1)}$$

2. The formula used to take account of the dependence of the electronic matrix element on the vibrational coordinate (i.e., which takes account of deviations from the Condon approximation) is

$$\sum_m \sqrt{(m+1)(m+2)\ldots(m+k)}\, J_{m,m+s} J_{m+k,m+s} q^{2m}$$

$$= \frac{x_0^k 2^{-k/2}}{(1-q^2)^{k+1}} \exp\left(-\frac{1+q^2}{1-q^2}\frac{x_0^2}{2}\right)$$

$$\times \sum_{l=0}^{k} q^{l-s}(-1)^l \binom{k}{l} I_{s-l}\left(\frac{q}{1-q^2}x_0^2\right), \quad q^2 < 1. \quad \text{(III.2)}$$

3. The formula for calculating the quasiline intensity for localized vibrations taking account of (in the first approximation) a small shift in the localized vibration frequency

---

* The formulas were obtained by R. A. Preém by the generating function method developed for the sums containing Franck–Condon integrals [73].

$$\sum_m J^2_{m,m+s}(\omega,\ \omega')\, q^{2m} = \frac{q^{-s}}{1-q^2}\exp\left(-\frac{1+q^2}{1-q^2}\frac{\bar{x}^2}{2}\right)\Bigg\{ I_s(Q)$$

$$+\frac{\bar{x}^2}{2}\frac{\omega-\omega'}{\omega}\left(\frac{q}{1-q^2}\right)^2 [I_{s+2}(Q)-2qI_{s+1}(Q)+2qI_{s-1}(Q)-I_{s-2}(Q)]\Bigg\}\cdot \text{(III.3)}$$

Here we have used the following notation: $J_{m,v}$ is the overlap integral for harmonic oscillator wave functions with unchanged frequency

$$J_{m,v}=\int \varphi_m(x)\,\varphi_v(x+x_0)\,dx,$$

where x_0 is the dimensionless shift in the equilibrium position ($x_0^2/2$ is the dimensionless Stokes loss), and $\varphi_m(x)$ and $\varphi_v(x)$ are defined by the equation $\varphi_m(x)=e^{-x^2/2}H_m(x)$, i.e., they differ from the wave functions defined by (II.52) by the factor $(\bar{q}_s)^{-1/2}$.

$J_{m,v}(\omega,\ \omega')$ is the same overlap integral but $\varphi_m(x)$ and $\varphi_v(x)$ belong to oscillators with the frequencies ω and ω' respectively. For a small change in the frequency the Stokes loss is

$$\bar{x}^2=x_0^2\frac{2\omega}{\omega+\omega'}.$$

The parameter q lies within the limits $1>q>0$. In the theory of quasiline spectra we have assumed the Boltzmann factor for this parameter, i.e.,

$$q=\exp(-\hbar\omega/2kT),$$

where ω is the oscillator frequency in the initial electronic state. It is not difficult to see that the average thermal energy $k\tau$ of the oscillator then enters into the argument of the exponential functions in Eqs. (III.1)–(III.3) ($k\tau/\hbar\omega = \frac{1}{2}\coth \hbar\omega/2kT$, see (7.17b)).

$I_s(y)$ is the modified Bessel function of the s-th order. This function can be defined from an infinite series (see, for example, [157])

$$I_s(y)=\sum_{k=0}^{\infty}\frac{1}{k!\,\Gamma(s+k+1)}\left(\frac{y}{2}\right)^{s+2k},$$

where $\Gamma(\nu)$ is the gamma function.

In Eq. (III.3) we have used the abbreviated notation

$$Q \equiv \bar{x}^2 \frac{q}{1 - q^2} .$$

4. Two formulas for the connection between overlap integrals encountered with linear and quadratic dependence of the electronic matrix element on the vibrational coordinate [73] are:

$$J_{m,v}^{(1)} \equiv \int \varphi_m(x) \varphi_v(x + x_0) x \, dx = \frac{1}{x_0} \left(v - m - \frac{x_0^2}{2} \right) J_{m,v}; \qquad \text{(III.4)}$$

$$J_{m,v}^{(2)} \equiv \int \varphi_m(x) \varphi_v(x + x_0) x^2 dx$$

$$= \frac{1}{x_0} \left[\frac{1}{x_0} \left(v - m - \frac{x_0^2}{2} \right) \left(v - m - \frac{x_0^2}{2} + 1 \right) J_{m,v} \right.$$

$$\left. - \sqrt{2} \sqrt{m + 1} \, J_{m+1,v} \right].$$

$$\text{(III.5)}$$

A Table of the Limiting Vibration Frequencies for Certain Ionic Crystals

Crystal	Vibrational branch	Limiting vibration frequency, cm^{-1}		Remarks
		$k = 0$	$k = k_{max}$ [100]	
NaF	TA	—	143 (T)	(T) theory considers the
	LA	—	234 (T)	deformability of the
	TO	202 (T); 247 (IR)	218 (T)	ions [63];
	LO	442 (T)	280 (T)	(IR) from [202].
NaCl	TA	—	83.6 (T)	(T) from [203];
	LA	—	141 (T)	(IR) from [202, 204]
	TO	162 (T); 164 (IR)	—	
	LO	262 (T)	—	
NaBr	TA	—	52 (T)	(T) from [63];
	LA	—	83 (T)	(IR) from [202]
	TO	143 (T); 134 (IR)	161 (T)	
	LO	207 (T)	190 (T)	
NaI	TA	—	37.3 (T); 40.8 (N)	(T) from [63];
	LA	—	56 (T); 61 (N)	(N) from slow neutron
	TO	118 (T); 117 (IR) 120 (N)	129 (T); 127 (N)	scattering at 110 K [205]
	LO	176 (T); 167 (N)	145 (T); 131 (N)	

continued:

Crystal	Vibrational branch	Limiting vibration frequency, cm^{-1}		Remarks
		$k = 0$	$k = k_{max}$ [100]	
KF	TA	—	91 (T)	(T) from [63]
	LA	—	141 (T)	
	TO	147 (T)	149 (T)	
	LO	333 (T)	185 (T)	
KCl	TA	—	64 (T)	(T) from [203];
	LA	—	116 (T)	(IR) from [204];
	TO	135 (T); 134 (RS) 138 (RS); 142 (IR)	142 (T)	(RS) from [206, 208]
	LO	213	154 (T)	
KBr	TA	—	45.2 (T); 41.5 (T); 44 (IR)	(T) from [63]; (N) from slow neutron scattering at 90 K [207]; (IR) from [202, 155]; (IR) from [206, 208]. The frequency spectrum contains a gap between 95 and 100 cm^{-1} [63, 205].
	LA	—	75 (T); 72 (N); 79.5 (IR)	
	TO	108 (T); 113 (IR); 110 (RS); 120 (N); 126 (RS)	116 (T); 125 (N)	
	LO	157 (T); 167 (N); 168 (RS)	133 (T); 134 (N)	
KI	TA	—	32 (T)	(T) from [63]; (IR) from [202]; (RS) from [206]. The frequency spectrum contains a gap from 65 to 88 cm^{-1} [63] or from 69 to 87 cm^{-1} [155].
	LA	—	53 (T)	
	TO	98 (T); 98 (IR); 100 (RS)	108 (T)	
	LO	132 (T)	124 (T)	
CaF$_2$	TA	—	318 (T)	(IR) from [209]; (T) from [210].
	LA	—	175 (T)	
	TO	{ 257 (IR) 257 (IR)	{ 190 (T) 314 (T)	
	LO	463 (IR); 441 (T)	327; 328 (IR)	
	TR	{ 321 (T) 321 (T)	{ 212 (T) 346 (T)	
	LR	321 (T)	152 (T)	

continued:

Crystal	Vibrational branch	Limiting vibration frequency, cm^{-1}		Remarks
		k = 0	k = k$_{max}$ [100]	
SrF$_2$	TO	217 (IR)	—	(IR) from [209]
	LO	374 (IR)	316 (IR)	
	ac.	—	99 (IR)	
BaF$_2$	TO	184 (IR)	—	(IR) from [209]
	LO	326 (IR)	278 (IR)	
	ac.	—	94 (IR)	
CaO	—	—	365 *	(*) from [48]
SrO	—	—	213 *	
MgO	opt.	490 (L)	578 *	(L) from vibrational
		430 (L)		structure in the lumines-
	ac.	270 (L)		cence spectrum of V^{2+}
				centers in MgO [211]

The table gives values of the limiting frequencies in the phonon spectrum for alkali halide and certain other ionic crystals.
The sources of the information are papers on infrared absorption
(IR), or on Raman scattering (RS). The most complete information on the phonon spectrum comes from experiments on slow neu-
tron scattering (N). Some information on the vibrational spectrum
of the host crystal can also be obtained from low-temperature
luminescence and absorption spectra of localized centers in the
crystals (L).

The table also gives limiting frequencies obtained from theoretical calculations using polarizable ion models (T). The original papers are cited under Remarks. We use the following notation for the branches: T, transverse; L, longitudinal; A, acoustic; O, optical; R, Raman.

The frequency of the longitudinal optical branch at k = 0 (LO,
k = 0) is usually the highest frequency in the crystal.

The Proportionality Coefficient between the Transition Probability and the Square of the Matrix Element of the Perturbation Operator in the Quantum Theory of Radiation

In the quantum theory of radiation the photon radiation and absorption probabilities for the system under consideration are given in the nonrelativistic approximation by the following expressions [90]

$$W_r(\omega) = \frac{\omega}{2\pi\hbar c^3}(\bar{n}+1)\left|\left\langle \sum_l \frac{Q_l}{m_l}(\varepsilon p_l)\exp(-iKr_l)\right\rangle_{fi}\right|^2, \qquad (V.1)$$

$$W_a(\omega) = \frac{\omega}{2\pi\hbar c^3}\bar{n}\left|\left\langle \sum_l \frac{Q_l}{m_l}(\varepsilon p_l)\exp(-iKr_l)\right\rangle_{fi}\right|^2. \qquad (V.2)$$

Here $W_r(\omega)$ is the probability per unit time that the system radiates a photon of frequency ω (per unit frequency range), wave vector K (per unit solid angle) $(|K|=\omega/c)$, and of polarization ε; $W_a(\omega)$ is the probability that the system absorbs one photon of frequency ω per unit time; Q_l is the charge, m_l is the mass, r_l is the position vector, and p_l is the momentum operator of the l-th particle of the system; $\bar{n}$ is the average number of photons present before the radiation (absorption) in the field oscillator of frequency ω per unit frequency range having wave vector K (per unit solid angle) and polarization ε ; and i and f are the numbers of the initial and final states of the system.

In the dipole approximation, which is valid when the light wavelength is greater than the size of the system (the atom or the impurity center), we have $|Kr_l| = \bar{r}_l \lambda^{-1} \ll 1$, where $\bar{r}_l$ is the average size of the system. We can set $(-iKr_l) \approx 1$ and

$$\left\langle \sum_l \frac{Q_l}{m_l} (\varepsilon p_l) \exp(-iKr_l) \right\rangle_{fi} \to i \frac{E_f - E_i}{h} \langle \varepsilon D \rangle_{fi}, \tag{V.3}$$

where $D = \sum_l Q_l r_l$ is the dipole moment of the system, and E_i and E_f are the energies of the system in the states i and f.

Using the notation $|\langle \varepsilon D \rangle_{fi}|^2 = |P_{fi}|^2$ and introducing the radiation intensity $H = \frac{h\omega^3}{(2\pi)^3 c^2} \bar{n}$ instead of $\bar{n}$ we can express W_r and W_a as follows

$$W_r(\omega) = \left(\frac{4\pi^2 H(\omega)}{h^2 c} + \frac{\omega^3}{2\pi h c^3} \right) |P_{fi}|^2 = W_r^{ind} + W_r^{sp}, \tag{V.4}$$

$$W_a(\omega) = \frac{4\pi^2 H(\omega)}{h^2 c} |P_{fi}|^2, \tag{V.5}$$

where the intensity $H(\omega)$ of the radiation denotes the radiated energy (with frequency ω per unit frequency range, wave vector K per unit solid angle, and with polarization ε) passing through unit area per unit time. The first term in the radiation probability is interpreted as induced emission W_r^{ind}, and the second as spontaneous emission W_r^{sp}.

The corresponding radiated power S_r and absorbed power S_a are
$$S_r = h\omega W_r, \quad S_a = h\omega W_a.$$

Since the radiation and absorption by the impurity atom (ion) occur within a crystal medium, we add correction factors to the expressions for the probabilities (see, for example, [9, 84]):

$$S_r^{sp} = \frac{n^3}{\varepsilon} \left(\frac{\mathscr{E}_e}{\mathscr{E}} \right)^2 \frac{\omega^4}{2\pi c^3} |P_{fi}|^2, \tag{V.6}$$

$$S_a = \frac{n}{\varepsilon} \left(\frac{\mathcal{E}_e}{\mathcal{E}} \right)^2 \frac{4 \pi^2 H(\omega) \omega}{hc} |P_{fi}|^2, \tag{V.7}$$

where n is the refractive index, ε is the dielectric constant at optical frequencies, $\mathcal{E}_e$ is the acting field in the medium and $\mathcal{E}$ is the average macroscopic electric field. All of these depend, in general, on the frequency ω.

Thus the spontaneous emission power $S_r^{sp}(\omega)$ and the absorption coefficient $\varkappa(\omega) = S_a(\omega) H(\omega)^{-1}$ contain explicit frequency dependences of the form ω^4 and ω respectively. The frequency dependence of $|P_{fi}(\omega)|^2$ is usually substantially stronger and it determines the shape of the vibronic band. Therefore we concentrate on this factor. However, in a detailed investigation of the band shapes one cannot forget about the additional frequency dependence in Eqs. (V.6) and (V.7). In nonlinear optical phenomena, which appear in a crystal under the action of a powerful beam of coherent laser light, these dependences play a substantial role.

If the given transition frequency ω corresponds to several different pairs of levels E_i and E_f, their contributions must be summed taking the weights of the various initial states into account,

$$I(\omega) = \frac{n^3}{\varepsilon} \left(\frac{\mathcal{E}_e}{\mathcal{E}} \right)^2 \frac{\omega^4}{2 \pi c^3} \sum_i n_i \sum_f |P_{fi}|^2 \delta (E_f - E_i - h\omega)$$

$$\approx A_r \sum_i n_i \sum_f |P_{fi}|^2 \delta (E_f - E_i - h\omega), \tag{V.8}$$

$$\varkappa(\omega) = \frac{n}{\varepsilon} \left(\frac{\mathcal{E}_e}{\mathcal{E}} \right)^2 \frac{4 \pi^2 \omega}{hc} \sum_i n_i \sum_f |P_{fi}|^2 \delta (E_f$$

$$- E_i - h\omega) \approx A_a \sum_i n_i \sum_f |P_{fi}|^2 \delta (E_f - E_i - h\omega), \tag{V.9}$$

where the functions $I(\omega)$ and $\varkappa(\omega)$ give the shapes of the emission and absorption spectra respectively, n_i is the number of systems in the i-th initial states, and the factors A_r and A_a are constants if we neglect the comparatively weak frequency dependence mentioned above. The delta function under the summation sign over final states f ensures the selection of only those transitions which correspond to frequency ω. If from the very beginning we consider only those terms which correspond to transitions with frequency ω in the sum over f, then the delta function need not be written.

In this approximation the shapes of the absorption and emission spectra are given by the same expression

This does not mean that ultimately the absorption and emission spectra always coincide; a substantial difference arises from the fact that the initial states are substantially different.

It is often convenient to write σ in a somewhat different way

$$\sigma(\omega) = A \sum_i v_i \sum_f |P_{fi}|^2 \delta(E_f - E_i - \hbar\omega). \qquad (V.10)$$

Here we have introduced the normalized occupation probability of the i-th state, $v_i = n_i N^{-1}$, where N is the total number of systems, and we have used a common notation for the constant factor $A =$ const. The description in Eq. (V.10) corresponds to the rule of "summing over final states and averaging over initial states."

Appendix VI

The Method of Moments

At high temperatures or for large Stokes losses the absorption
and emission spectra are characteristic bell-shaped curves (case
4, section 16). This shape is well described by a gaussian curve
multiplied by a correction function which corresponds to some
deviation of the shape from a precise gaussian curve, and the appearance of a small asymmetry. Similar shapes occur for many
distribution functions encountered in probability theory. The well-
known method of moments [212] is used for the quantitative characteristics of such distribution functions in probability theory.

The moment S_l of the l-th order for the distribution $\rho(x)$ is
given by the integral

$$S_l = \int x^l \rho(x)\, dx, \tag{VI.1}$$

where the integration runs over the whole range of x. Equation
(VI.1) gives the moments relative to the origin of coordinates.
Such moments are called, according to the terminology of probability theory, original moments. If the variable is discrete then the
integral is replaced by a sum

$$S_l = \sum_r x_r^l \rho_r. \tag{VI.2}$$

If for clarity we think of $\rho(x)$ as the mass distribution of
some body along the x axis, then S_0 gives the mass of the body.

The first moment is related to the coordinate $\bar{x}$ of the center of mass by

$$\bar{x} = \frac{S_1}{S_0}. \tag{VI.3}$$

The second moment S_2 is related to the moment of inertia.

If $\rho(x)$ is a bell-shaped curve then $\bar{x}$ is close to the coordinate of maximum $\rho(x)$ and

$$\bar{S}_2 \equiv S_2 - \frac{S_1^2}{S_0} = S_0(\bar{x^2} - (\bar{x})^2), \tag{VI.4}$$

i.e., the second moment relative to $\bar{x}$ ("The moment of inertia of the body relative to its center of mass") characterizes the width of the distribution. Similarly, $\bar{S}_3$ characterizes the asymmetry in $\rho(x)$, etc. The moments relative to $\bar{x}$ are called central moments. We will distinguish them by bars over the moment symbol.

The central moments $\bar{S}_l$ are expressed in terms of the original moments S_l by

$$\bar{S}_l = \sum_{p=0}^{l} (-1)^p \binom{l}{p}\left(\frac{S_1}{S_0}\right)^p S_{l-p}. \tag{VI.5}$$

For the first few moments we thus have

$$\bar{S}_0 = S_0, \quad \bar{S}_1 = 0, \quad \bar{S}_2 = S_2 - \frac{S_1^2}{S_0}, \quad \bar{S}_3 = S_3 - 3S_2\frac{S_1}{S_0} + 2\frac{S_1^3}{S_0^2}.$$

In describing a spectral band in a crystalline phosphor it is usual to give the position of its maximum and its half-width. A parameter characterizing the asymmetry is also introduced. Because the band is of nearly gaussian shape, several of its first few moments give as much information about the band as the position of the maxima, half-width, etc. which are usually given.

Theoretical calculation of the moments of the spectral curve is substantially simpler than calculation of the curve itself or the characteristics of it which are usually referred to. It has been shown by Lax [84] that one can obtain a number of quite general results for the moments of a vibronic spectral curve. In [106] the moments of spectra were calculated, taking into account anharmonicity of the vibrations and other correction factors. A comparison with experiment can also be carried out by solving the inverse problem: constructing $\rho(x)$ from its moments.

For this purpose it is appropriate to use Edgeworth's formula [84, 213), which allows us to construct $\rho(x)$ on the basis of the Gaussian curve and its derivatives, which have been tabulated [213].

$$\rho(x) = S_0\,\sigma^{-1}\left[\varphi(t) - \frac{\gamma_1}{3!}\varphi^{(3)}(t) + \frac{\gamma_2}{4!}\varphi^{(4)}(t) + \frac{10\,\gamma_1^2}{6!}\varphi^{(6)}(t) + \cdots\right],$$

where

(VI.6)

$$t \equiv \frac{x - \bar{x}}{\sigma}, \qquad \sigma^2 \equiv \overline{S}_2 S_0^{-1}, \qquad \gamma_1 \equiv \overline{S}_3\,(\overline{S}_2)^{-1/2}\,(S_0)^{1/2},$$

$$\gamma_2 \equiv \overline{S}_4\,(\overline{S}_2)^{-2}\,S_0 - 3, \qquad \varphi(t) \equiv (2\,\pi)^{-1/2}\exp(-t^2/2).$$

The first term in Eq. (VI.6) gives a gaussian curve. The second term gives $\rho(x)$ some asymmetry, whose magnitude depends on $\overline{S}_3$ through γ_1. The third and fourth terms are even and modify $\rho(x)$ around the maximum and in the wings.

Below we will give a simple derivation of the formula for the moments [85, 86].

The transition probability $W_{nn'}$ from state n to state n' is given in first-order perturbation theory by the absolute square of the matrix element $P_{nn'}$ of the perturbation operator $\hat{P}$ which gives rise to the transition

$$W_{nn'} = \frac{2\pi}{h}\,|P_{n'n}|^2, \qquad P_{n'n} \equiv \int \psi_{n'}^*\hat{P}\psi_n\,d\tau,$$

where ψ_n is the wave function of state n and the integration runs over all of the configuration space of the system.

For a continuous spectrum of states we have the corresponding transition probability density for $n \rightarrow n'$

$$W(n \rightarrow n') = \frac{2\pi}{h}\,|P_{n'n}|^2\,\rho(E_{n'}),$$

where $\rho(E_{n'})$ is the density of final states. In all of the expressions for the moments given below one must then transform from a summation to an integral over final states; $\rho(E_{n'})$ then enters as a factor (jacobian) in the transition to a continuous energy scale and the final result does not depend on whether the spectrum of states is continuous or discrete.

The l-th order moment of the transition probability distribution for $n \to n'$ in the transition energy scale (the factor $2\pi h^{-1}$ is dropped for simplicity) is given by

$$S_l = \sum_{n'} (E_{n'} - E_n)^l \, |P_{n'n}|^2, \qquad (VI.7)$$

where E_n is the energy of the n-th state.

We rewrite the power of the energy difference in the form of a binomial, expressing $|P_{n'n}|^2$ as the product of two integrals (matrix elements), and for the summation over n' we put in the corresponding hamiltonian operators in place of $E_{n'}$ and E_n in (VI.7). We thus find

$$S_l = \sum_{n'} \sum_{p=0}^{l} (-1)^p \binom{l}{p} \int E_{n'}^{l-p} \psi_n^* \hat{P} E_n^p \psi_n \, d\tau \int \psi_{n'} (\hat{P}\psi_n)^* \, d\tau'$$

$$= \sum_{n'} \sum_{p=0}^{l} (-1)^p \binom{l}{p} \int (H_{II}^{l-p} \psi_{n'})^* \, \hat{P} \, (H_I^p \psi_n) \, d\tau \int \psi_{n'} (\hat{P}\psi_n)^* \, d\tau', \qquad (VI.8)$$

where H_I and H_{II} are the hamiltonian operators of the initial and final states respectively ($H_I \psi_n = E_n \psi_n$ and $H_{II} \psi_{n'} = E_{n'} \psi_{n'}$). We do not initially assume that $H_I = H_{II}$ but retain the possibility of using different approximate hamiltonians for the different states of the system. This, in particular, is important when we subsequently use the adiabatic approximation.

Since the operator H_{II} is self-conjugate, Eq. (VI.8) can be written as

$$S_l = \sum_{n'} \sum_{p=0}^{l} (-1)^p \binom{l}{p} \int \psi_{n'}^* H_{II}^{l-p} \hat{P} H_I^p \psi_n \, d\tau \int \psi_{n'} (\hat{P}\psi_n)^* \, d\tau'.$$

Now using the completedness condition of the wave functions $\psi_{n'}$ of the final state, $\sum_{n'} \psi_{n'}^*(\tau)\psi_{n'}(\tau') = \delta(\tau - \tau')$, we can simply carry out a summation over n' which leads to the final expression for S_l:

$$S_l = \sum_{p=0}^{l} (-1)^p \binom{l}{p} \int (\hat{P}\psi_n)^* H_{II}^{l-p} \hat{P} H_I^p \psi_n \, d\tau = \sum_{p=0}^{l} (-1)^p \binom{l}{p} \int \psi_n^* \hat{P}^+ H_{II}^{l-p} \hat{P} H_I^p \psi_n \, d\tau,$$

$$(VI.9)$$

where $\hat{P}^+$ is conjugate to $\hat{P}$.

It is important that the formula obtained does not now contain the wave functions ψ_n of the final state. This is one of the advantages of the method of moments.

In considering vibronic transitions in molecules or luminescent centers in crystalline phosphors, we introduce the dipole and adiabatic approximations, i.e., $\hat{P} = D$, where D is the projection of the dipole moment of the system along the direction of the electric vector of the light wave, and the wave function is the product of electronic and vibrational wave functions

$$\psi_n(r,\ R) = \Phi_l(r,\ R)\varphi_{li}(R),\ \ \psi_{n'}(r,\ R) = \Phi_m(r,\ R)\varphi_{mf}(R).$$

Here l and m are the indices of the electronic states and i and f are the indices of the vibrational states (see section 4). We are interested in the distribution of intensity in the vibrational structure of a particular electronic transition $l \to m$. Therefore we integrate over the electronic coordinates and introduce the electronic matrix elements $D_{ml}(R)$ which act as multiplicative operators on the vibrational wave functions.

We now bring in

$$P_{mf,li} = \int \varphi_{mf}^*(R)\, D_{ml}(R)\, \varphi_{li}(R)\, dR,$$

as the matrix element of the perturbation operator, where $D_{ml}(R)$ is the matrix element of the dipole electronic transition (4.2a).

Equation (VI.9) takes the form

$$S_l = \sum_{p=0}^{r} (-1)^p \binom{r}{p} \int \varphi_{li}^*(R)\, D_{ml}^*(R)\, H_{II}^{r-p} D_{ml}(R)\, H_I^p \varphi_{li}(R)\, dR. \tag{VI.10}$$

with these approximations.

We write the formulas obtained for the moments S_r of orders $r = 0, 1, 2, 3$. We also average over the vibrational levels of the initial electronic state, for which we introduce the normalized occupation probabilities of the levels, $v_i = n_i N^{-1}$. We then find

$$S_0 = \sum_i v_i \int |\varphi_{li}(R)|^2 |D_{ml}(R)|^2\, dR, \tag{VI.11a}$$

$$S_1 = \sum_i v_i \int \varphi_{li}^*(R)\, \{D_{ml}^*(R)\, H_{II} D_{ml}(R) - |D_{ml}(R)|^2 H_I\} \varphi_{li}(R)\, dR, \tag{VI.11b}$$

$$S_2 = \sum_i v_i \int \varphi_{li}^* (R) \{ D_{ml}^* (R) H_{II}^2 D_{ml} (R)$$

$$- 2D_{ml}^* (R) H_{II} D_{ml} (R) H_I + |D_{ml}(R)|^2 H_I^2 \} \varphi_{li} (R) dR, \qquad \text{(VI.11c)}$$

$$S_3 = \sum_i v_i \int \varphi_{li}^* (R) \{ D_{ml}^* (R) H_{II}^3 D_{ml}(R)$$

$$- 3D_{ml}^* (R) H_{II}^2 D_{ml} (R) H_I + 3D_{ml}^* (R) H_{II} D_{ml}(R) H_I^2$$

$$- |D_{ml} (R)|^2 H_I^3 \} \varphi_{li} (R) dR. \qquad \text{(VI.11d)}$$

Equations (VI.11) agree with the equations of Lax [84]. Setting $D_{ml}(R) = D = const$ in these equations, we arrive at the Lax equations in the Condon approximation

$$S_0 = |D|^2 \int \rho (R) dR = |D|^2, \qquad \text{(VI.12a)}$$

$$S_1 = |D|^2 \sum_i v_i \int \varphi_{li}^* (R)(H_{II} - H_I) \varphi_{li} (R) dR, \qquad \text{(VI.12b)}$$

$$S_2 = |D|^2 \sum_i v_i \int \varphi_{li}^* (R)(H_{II}^2 - 2H_{II}H_I + H_I^2) \varphi_{li} (R) dR, \qquad \text{(VI.12c)}$$

$$S_3 = |D|^2 \sum_i v_i \int \varphi_{li}^* (R)(H_{II}^3 - 3H_{II}^2 H_I + 3H_{II}H_I^2$$

$$- H_I^3) \varphi_{li} (R) dR, \qquad \text{(VI.12d)}$$

where

$$\rho (R) \equiv \sum_i v_i |\varphi_{li} (R)|^2 \qquad \text{(VI.13)}$$

is the quantum-mechanical coordinate distribution function in the initial electronic state.

The equations for the zeroth moment, (VI.11a) and (VI.12a), within the normalization of the distribution function, agree with (9.2) and (9.3), respectively.

In a number of cases the integrands in (VI.12b, c, and d) transform such that they reduce to a product of $\rho(R)$ with a certain simple function (often a polynomial) of R.

If the vibrations in the initial electronic state are described in the harmonic approximation (in the final electronic state they can be anharmonic!) then the quantum-mechanical coordinate distribution function in thermal equilibrium, $\rho(R, T)$ is easily calculated and assumes a simple explicit form

$$\rho(R,\,T) = \sum_{i} v_i |\varphi_{1i}(R)|^2 = \frac{1}{\varepsilon\sqrt{\pi}}\, e^{-R^2/\varepsilon^2}, \qquad \text{(VI.14)}$$

where $\varepsilon^2 \equiv 2kT/m\omega^2$, m and ω are the mass and frequency of the oscillator, and $kT \equiv (1/2)\hbar\omega \operatorname{cth}(\hbar\omega/2kT)$. After substituting $\rho(R, T)$ in the form (VI.14) into (VI.12) the moments can be expressed in terms of easily calculated integrals which are often encountered in statistical physics.

In deriving Eq. (VI.9), we have not made any special assumptions and it is widely applicable. In particular, Eq. (VI.9) contains the known spectroscopic sum rules and is a general expression for obtaining new ones [69].

If the spectrum has complicated structure it is practically impossible to deduce a detailed intensity distribution from the moments. In this case the significance of the moments as spectroscopic sum rules remains. Equation (VI.9) can be written as a trace (sum over all diagonal elements) of the matrix $\hat{\rho}\hat{S}_l$

$$S_l = \operatorname{Sp}(\hat{\rho}\hat{S}_l),$$

where $\hat{\rho}$ is the density matrix and $\hat{S}_l$ is the moment operator defined by*

$$\hat{S}_l \equiv \sum_{p=0}^{l} (-1)^p \binom{l}{p} \hat{P}^+ H_{\mathrm{II}}^{l-p} \hat{P} H_{\mathrm{I}}^p. \qquad \text{(VI.15)}$$

For a system in thermal equilibrium the density matrix takes the form

$$\hat{\rho} = \exp(-H_{\mathrm{I}}/kT)\{\operatorname{Sp}[\exp(-H_{\mathrm{I}}/kT)]\}^{-1}, \qquad \text{(VI.16)}$$

where H_{I} is the Hamiltonian operator of the system in the initial state.

The trace of a matrix, as is well known, does not depend on what complete system of functions is used to calculate it (see, for

---

* From the derivation of Eq. (VI.8) it is not difficult to see that the definition of the moment operator is not unique. For moments of time-dependent spectra it is necessary to define it in Hermitian form [214]

$$\hat{S}_l \equiv \sum_{p=0}^{l} (-1)^p \binom{l}{p} \{\hat{P}^+ H_{\mathrm{II}}^{l-p} \hat{P} H_{\mathrm{I}}^p + \hat{P} H_{\mathrm{I}}^p \hat{P}^+ H_{\mathrm{II}}^{l-p}\}.$$

example, [70, 104]). This means that it is not necessary to know the wave functions ψ_n of the initial state to find the moments from Eq. (VI.9a).

This is the great advantage of the method of moments in problems involving averaging over initial states of the system.

Exercise 42. Calculate S_l for dipole and quadrupole absorption spectra in vibrational transitions of a harmonic oscillator.

The Adiabatic Approximation
and the Variational Principle

VII.1 Derivation of the Adiabatic Approximation from the Variational Principle*

The adiabatic approximation contains two elements: 1) postulation, and 2) the requirement of minimum energy (variational principle).

In the first place, proceeding from clear physical ideas we postulate that a fairly good approximation to the wave function can be had from a function of the form (see (1.3a))

$$\psi_{li}(r, R) = \phi_l(r, R)\, \varphi_{li}(R), \tag{VII.1}$$

where the function ϕ_l is defined as a normalized eigenfunction of the equation

$$\left(H + \frac{h^2}{2} \sum_\alpha \frac{\Delta_\alpha}{M_\alpha}\right) \phi_l(r, R) = W_l(R)\, \phi_l(r, R), \tag{VII.2}$$

$$\int \phi_l^2\, dr = 1. \tag{VII.3}$$

---

* A discussion of the variational principle can be found in almost any textbook on quantum mechanics (see, for example, [70]).

We let H be the exact hamiltonian (1.1) so that (VII.2) agrees with Eq. (1.6a). We will also assume that the functions ϕ and φ are real, which is always possible when there is no magnetic field.

The problem now consists of finding the equation which should be satisfied by the functions $\varphi_{li}(R)$ so that among all possible functions defined by Eqs. (VII.1)-(VII.3) we find the "best" approximate solution of the Schrödinger equation (1.2). The "best" approximate solution (within the class of competing functions) is to be chosen by the variational principle — the requirement that the average value of the total energy of the system be a minimum.

Mathematically this means finding the minimum value of the integral

$$\int \phi_l \varphi_{li} H \phi_l \varphi_{li} \, dr \, dR \equiv \bar{H} \tag{VII.4}$$

as a function of the form of $\varphi_{li}(R)$, taking (VII.2) and (VII.3) into account, with the additional condition (normalization)

$$\int \varphi_{li}^2(R) \, dR = 1. \tag{VII.5}$$

Integrating over r and considering (VII.2, 3), we can transform (VII.4) as follows

$$\bar{H} = \int \varphi_{li}(R) \left[W_l(R) - \frac{h^2}{2} \sum_\alpha \frac{\varDelta_\alpha}{M_\alpha} - \frac{h^2}{2} \sum_\alpha \frac{1}{M_\alpha} \int \phi_l \varDelta_\alpha \phi_l \, dr \right] \varphi_{li}(R) \, dR. \tag{VII.6}$$

Carrying out a variation of (VII.6) with the side condition (VII.5) on the variation of the function φ_{li}, we find the following equation for φ_{li}

$$\left[-\frac{h^2}{2} \sum_\alpha \frac{\varDelta_\alpha}{M_\alpha} + W_l(R) - \frac{h^2}{2} \sum_\alpha \frac{1}{M_\alpha} \int \phi_l \varDelta_\alpha \phi_l \, dr \right] \varphi_{li}(R) = E_{li} \varphi_{li}(R), \tag{VII.7}$$

whose eigenvalue E_{li} is to be minimized within the class of functions (VII.1)-(VII.3) and (VII.5) and is the average energy of the whole system. (Formally E_{li} appears in (VII.7) as a Lagrange multiplier corresponding to the condition (VII.5)).

Equation (VII.7) determines the set of orthonormalized vibrational functions which correspond to a given electronic state l in the adiabatic approximation. Essentially the variational derivation of the vibrational equation leads to a refinement of Eq. (1.7a);

we find $U_l(R)$ defined by Eq. (1.8) for the potential energy operator.

Since the functions (VII.1) are not exact eigenfunctions of the hamiltonian H defined by (1.1), it follows from the variational principle that $E'_{00} \geq E_0$, where E_0 is the exact lowest eigenvalue of H and E'_{00} is the approximate eigenvalue from the improved adiabatic approximation, i.e., the value of the integral $\overline{H}$ in (VII.4) if we use the solutions ϕ_0 and φ_{00} of Eqs. (VII.2) and (VII.7) for the functions.

VII.2   Inequalities for the
Energy Eigenvalues

We will now show that the lowest energy eigenvalue E_{li} ($l = 0$, $i = 0$) calculated in the u s u a l adiabatic approximation, i.e., according to (1.3a), (1.6a), and (1.7a), lies l o w e r than the lowest energy level of the exact hamiltonian H (1.1). This problem was first solved by Bratsev [215]. We give a simple demonstration, following Epstein [216].

We write the exact hamiltonian (1.1), the exact Schrödinger equation, and the usual adiabatic approximation (equations (1.6a) and (1.7a)) in the form

$$H = T_r + T_R + V(r, R), \qquad H\psi_0 = E_0\psi_0 ; \qquad \text{(VII.8)}$$

$$[T_r + V(r, R)] \phi_0 = W_0(R) \phi_0 ; \qquad \text{(VII.9)}$$

$$[T_R + W_0(R)] \varphi_{00} = E_{00}\varphi_{00} , \qquad \text{(VII.10)}$$

where T_r and T_R are kinetic energy operators for the electrons and nuclei respectively, and E_0 and E_{00} are the energies of the lowest level of the exact spectrum and the lowest vibronic (rotational) level in the usual adiabatic approximation, respectively. The wave functions ψ_0, ϕ_0, and φ_{00} are normalized to unity.

We want to show that

$$E_0 \geq E_{00} . \qquad \text{(VII.11)}$$

The derivation is based on the fact that ψ_0, ϕ_0, and φ_{00} are exact eigenfunctions of Eqs. (VII.8)–(VII.10), respectively. Therefore we have

$$\int \psi_0^* H\psi_0 \, dr \, dR = E_0 , \qquad \text{(VII.12)}$$

$$\int \phi_0^*[T_r + V(r, R)] \phi_0 \, dr = W_0(R), \tag{VII.13}$$

$$\int \varphi_{00}^*[T_R + W_0(R)] \varphi_{00} \, dR = E_{00}. \tag{VII.14}$$

The integrals on the left-hand sides of these equations are essentially the average values of the energy for the corresponding hamiltonians. According to the variational principle, these integrals increase if we replace the eigenfunction (for the lowest state) of the hamiltonian by any other function.

Replacing ϕ_0 by ψ_0 on the left-hand side of Eq. (VII.13), we find

$$\int \psi_0^*[T_r + V(r, R)] \psi_0 \, dr \geqslant W_0(R) \int \psi_0^* \psi_0 \, dr. \tag{VII.15}$$

(The equality occurs if ψ_0 also turns out to be a ground-state eigenfunction of the hamiltonian $T_r + V(r, R)$.)

We write (VII.12) similarly and take (VII.15) into account, finding

$$E_0 = \int \psi_0^*[T_r + T_R + V(r, R)] \psi_0 \, dr \, dR$$

$$= \int \psi_0^*[T_r + V(r, R)] \psi_0 \, dr \, dR + \int \psi_0^* T_R \psi_0 \, dr \, dR$$

$$\geqslant \int \psi_0^* W_0(R) \psi_0 \, dr \, dR + \int \psi_0^* T_R \psi_0 \, dr \, dR$$

$$= \int dr \int dR \, \psi_0^*[T_R + W_0(R)] \psi_0. \tag{VII.16}$$

The integral over R in the latter expression is the integral (VII.14) in which the exact eigenfunction φ_{00} is replaced by $\psi_0(r, R)$ (r is regarded as a parameter here). It then follows from the variational principle that

$$\int \psi_0^*[T_R + W_0(R)] \psi_0 \, dR \geqslant E_{00} \int \psi_0^* \psi_0 \, dR. \tag{VII.17}$$

The inequality (VII.16) can now be written as

$$E_0 \geqslant E_{00} \int \psi_0^*(r, R) \psi_0(r, R) \, dr \, dR. \tag{VII.18}$$

Equation (VII.11) also follows from this, taking the normalization into account.

Thus the lowest energy level E_{00} calculated in the u s u a l adiabatic approximation is always l o w e r than the exact lowest energy level E_0. The usual adiabatic approximation is given by Eqs. (1.3), (1.6), and (1.7) or (1.3a), (1.6a), and (1.7a).

The improved adiabatic approximation differs from the usual replacement of the nuclear potential energy operator $W_l(R)$ in Eq. (1.7a) by $U_l(R)$ from (1.8), or in other words, the replacement of Eq. (1.7a) by (VII.7). From the variational derivation of the improved adiabatic approximation given in section 1 of this Appendix it follows that the lowest level E'_{00} calculated using it always lies lower than the lowest exact energy level E_0. Finally we have

$$E'_{00} \geqslant E_0 \geqslant E_{00}. \tag{VII.19}$$

VII.3   The Hellmann – Feynman Theorem

This theorem asserts that if the hamiltonian $H(\lambda)$ depends on a parameter λ there is an equality between the λ derivative of the energy eigenvalue and the expectation value of the λ derivative of the potential energy operator [217, 218]

$$\frac{\partial E}{\partial \lambda} = \int \psi^* \frac{\partial V}{\partial \lambda} \psi \, d\tau, \tag{VII.20}$$

where $\psi = \psi(\tau, \lambda)$ is an eigenfunction of the hamiltonian $H(\lambda)$ and τ denotes the set of dynamical variables.

If we understand the coordinates R of the heavy particles for λ and we take the hamiltonian of the electronic equation (1.6a) (or VII.9) for $H(R)$, then the Hellmann–Feynman theorem gives the following expression for the R derivative of the adiabatic potential

$$\frac{\partial W_l(R)}{\partial R} = \int \phi_l^* \frac{\partial V(r, R)}{\partial R} \phi_l \, dr. \tag{VII.21}$$

We will prove this relation. We write the eigenvalue $W_l(R)$ in integral form (see VII.13) and calculate the derivative with respect to R.

We have

$$W_l(R) = \int \phi_l^*(r, R) H(R) \phi_l(r, R) \, dr, \tag{VII.22}$$

$$\frac{\partial W_l}{\partial R} = \int \phi_i^* \frac{\partial H}{\partial R} \phi_l \, dr + \int \frac{\partial \phi_i^*}{\partial R} H\phi_l \, dr + \int \phi_i^* H \frac{\partial \phi_l}{\partial R} \, dr$$

$$= \int \phi_i^* \frac{\partial H}{\partial R} \phi_l \, dr + W_l \int \frac{\partial \phi_i^*}{\partial R} \phi_l \, dr + W_l \int \phi_i^* \frac{\partial \phi_l}{\partial R} \, dr$$

$$= \int \phi_i^* \frac{\partial H}{\partial R} \phi_l \, dr + W_l \frac{\partial}{\partial R} \int \phi_i^* \phi_l \, dr. \tag{VII.23}$$

Here we have used the fact that the operator H is self-conjugate

$$\int \phi_i^* H \frac{\partial \phi_l}{\partial R} \, dr = \int \frac{\partial \phi_l}{\partial R} H\phi_i^* \, dr. \tag{VII.24}$$

and $H\phi_l = W_l \phi_l$. It is easily seen that the latter term on the right-
hand side of (VII.23) is zero; the normalization integral is unity
and does not depend on R. If we also take into account the fact that
the kinetic energy operator in the hamiltonian H does not depend on
the coordinate R, we have

$$\frac{\partial H}{\partial R} = \frac{\partial T_r}{\partial R} + \frac{\partial V(r, R)}{\partial R} = \frac{\partial V(r, R)}{\partial R}. \tag{VII.25}$$

Equation (VII.21) follows from (VII.23). The generalization
of the theorem to that case of N independent parameters (coordinates) $R_1, R_2, \ldots, R_N$ is trivial.

See [219–223], for example, for the application of the
Hellmann–Feynman theorem to the particular problem of molecular theory.

References

1. J. Franck, Trans. Faraday Soc., 21:536 (1925).
2. E. Condon, Phys. Rev., 32:858 (1928).
3. G. Herzberg, Molecular Spectra and Molecular Structure, Vol. 1: Spectra of Diatomic Molecules, Van Nostrand, New York (1950).
4. M. A. El'yashevich, Atomic and Molecular Spectroscopy, Fizmatgiz (1962).
5. S. I. Vavilov, Collected Works, Vol. 1, Paper No. 7. Izd. Akad. Nauk SSSR (1954); Phil. Mag., 43:307 (1922).
6. B. S. Neporent, Zh. Fiz. Khim., 30:1048 (1956); Izv.. Akad. Nauk SSSR, ser. fiz., 22:1372 (1958).
7. B. I. Stepanov, Luminescence of Complex Molecules, Izd. Akad. Nauk Byeloruss. SSR (1955).
8. B. I. Stepanov and V. P. Gribkovskii, Introduction to the Theory of Luminescence, Izd. Akad. Nauk Byeloruss. SSR (1963).
9. S. I. Pekar, Zh. Éksp. Teor. Fiz., 20:510 (1950); Usp. Fiz. Nauk, 50:197 (1953).
10. S. I. Pekar, Studies of the Electron Theory of Crystals, Gostekhizdat (1951).
11. K. Huang and A. Rhys, Proc. Roy. Soc., A204:406 (1950).
12. A. S. Davydov, Zh. Éksp. Teor. Fiz., 24:197 (1953).
13. M. A. Krivoglaz and S. I. Pekar, Trudy Inst. Fiziki Akad. Nauk Ukr. SSR, No. 4:37 (1953).
14. J. Frenkel, Phys. Rev., 37:17 (1931); Collected Selected Works, Vol. II, Izd. Akad. Nauk SSSR (1958), p. 158.
15. V. L. Levshin, Zh. Fiz. Khim., 2:641 (1931); Photoluminescence of Liquids and Solids, Gostekhizdat (1951); Z. Phys., 72:368, 382 (1931).
16. D. I. Blokhintsev, Zh. Éksp. Teor. Fiz., 9:459 (1939); J. Phys. USSR, 1:117 (1931).
17. W. E. Lamb, Phys. Rev., 55:190 (1939) (see also A. Akhiezer and I. Pomeranchuk, Some Problems of Nuclear Theory, Gostekhizdat (1948)).
18. I. V. Obreimov, Zh. Russ. Fiz. Khim. Obshch., 59:548 (1927); I. V. Obreimov and A. F. Prikhot'ko, Phys. Zs. Sow., 1:203 (1932).
19. J. Becquerel, Le Radium, 4:49 (1907); 5:5 (1908); 6:327 (1909).
20. P. P. Feofilov, Proceedings of the Fifth Conference on Luminescence, Tartu (1957), p. 3.
21. M. A. El'yashevich, Rare Earth Spectra, Gostekhizdat (1953).

22. É. V. Shpol'skii, Usp. Fiz. Nauk, 71:215 (1960); 77:32 (1962); 80:255 (1963).
23. A. N. Sevchenko and B. I. Stepanov, Zh. Éksp. Teor. Fiz., 21:212 (1951).
24. L. F. Johnson, R. E. Dietz, and H. J. Guggenheim, Phys. Rev. Letters, 11:318 (1963).
25. D. Mergerian and J. Markham, Advances in Quantum Electronics, ed. J. R. Singer, New York—London (1961), p. 267.
26. E. D. Trifonov, Dokl. Akad. Nauk SSSR, 147:826 (1962).
27. K. K. Rebane and V. V. Khizhnyakov, Optika i Spektr., 14:491 (1963).
28. R. H. Silsbee and B. D. Fitchen, Revs. Mod. Phys., 36:433 (1964).
29. E. F. Gross, B. S. Razbirin, and S. A. Permogorov, Dokl. Akad. Nauk SSSR, 147:338 (1962).
30. E. F. Gross, B. S. Razbirin, and S. A. Permogorov, Dokl. Akad. Nauk. SSSR, 154:1306 (1964).
31. Yu. Kagan and Ya. A. Iosilevskii, Zh. Eksp. Teor. Fiz., 42:259 (1962); 44:284 (1963).
32. R. Brout and W. Visscher, Phys. Rev. Lett., 9:54 (1962).
33. P. Dawber and R. Elliot, Proc. Roy. Soc., A273:222 (1963); Proc. Phys. Soc., 81:458 (1963).
34. A. A. Maradudin, Revs. Mod. Phys., 36:417 (1964).
35. W. Visscher, Phys. Rev., 129:28 (1963).
36. A. A. Maradudin, Repts. Prog. Phys., 28:331 (1965); Scientific Paper 65-9F5-442-P5, Westinghouse Research Laboratories, Pittsburgh (1965).
37. Yu. M. Kagan, Zh. Éksp. Teor. Fiz., 47:7 (1964).
38. V. V. Khizhnyakov, Izv. Akad. Nauk Eston. SSR, ser. fiz.-mat. i tekhn. nauk, 14:94 (1965).
39. M. A. Krivoglaz, Fiz. Tverd. Tela, 6:1707 (1964).
40. K. K. Rebane, N. N. Kristofel', E. D. Trifonov, and V. V. Khizhnyakov, Izv. Akad. Nauk Eston. SSR, ser. fiz.-mat. i tekhn. nauk, 13:87 (1964).
41. Crystal Spectroscopy, Proceedings of the Symposium on Spectroscopy of Crystals Containing Rare-Earth Elements and Iron-Group Elements, Feb. 3-6, 1965, Moscow, Izd. Nauka (1966).
42. J. J. Hopfield, Proc. Internat. Conf. Semicond. Phys., Exeter (1962), p. 75.
43. D. Fitchen, R. Silsbee, T. Fulton, and E. Wolff, Phys. Rev. Letters, 11:275 (1963).
44. C. B. Pierce, Phys. Rev., 135:A83 (1964).
45. D. B. Fitchen, H. R. Fetterman, and C. B. Pierce, Solid State Communications, 4:205 (1966).
46. D. B. Fitchen, in: Physics of Solids at High Pressures, ed. C. T. Tomizuka and R. M. Emrick, Academic Press, New York—London (1965), p. 383.
47. D. B. Fitchen, Preprints of the International Conference on Luminescence, 5:57, Budapest (1966).
48. Max Born and Kun Huang, Dynamical Theory of Crystal Lattices, Oxford Univ. Press, New York (1954).
49. É. V. Shpol'skii, Atomic Physics, Vol. 1, Fizmatgiz (1963).
50. É. I. Adirovich, Some Problems of the Theory of Crystal Luminescence, Gostekhizdat (1951).
51. K. K. Rebane and O. I. Sil'd, Izv. Akad. Nauk Eston. SSR, ser. fiz.-mat. i tekhn. nauk, 13:165 (1964).

52. M. G. Veselov, Elementary Quantum Theory of Atoms and Molecules, Gostekhiz-dat (1955).

53. N. N. Kristofel', Trudy Inst. Fiz. Astron., Akad. Nauk Eston. SSR, No. 10:3 (1959); No. 11:180 (1960); N. N. Kristofel', Izv. Akad. Nauk SSSR, ser. fiz., 24:533 (1961); N. N. Kristofel', Optika i Spektr., 9:615 (1960).

54. N. N. Kristofel' (N. N. Kristoffel), Zur Physik und Chemie der Kristallphosphore, II, Berlin (1962), p. 5.

55. N. D. Potekhina, Optika i Spektr., 8:832 (1960); A. A. Kiselev and I. V. Abaren-kov, Optika i Spektr., 9:765 (1960); A. Gold, J. Phys. Chem. Solids, 18:218 (1961).

56. E. Wilson, J. Decius, and P. Cross, Molecular Vibrations: Theory of Infrared and Raman Vibrational Spectra, McGraw-Hill, New York (1955).

57. A. I. Ansel'm, Introduction to the Theory of Semiconductors, Fizmatgiz (1962).

58. I. M. Lifshits, Zh. Éksp. Teor. Fiz., 17:1017, 1076 (1947); I. M. Lifshits (I. M. Lifsic), Nuovo Cimento, Vol. 3, Suppl. 4:716 (1956).

59. E. Montroll and R. Potts, Phys. Rev., 100:525 (1955).

60. G. S. Zavt, Fiz. Tverd. Tela, 7:2109 (1965).

61. G. S. Zavt, Fiz. Tverd. Tela, 5:1946 (1963); Trudy Inst. Fiz. Astron. Akad. Nauk Eston. SSR, No. 27:69 (1964).

62. G. S. Zavt and N. N. Kristofel', Fiz. Tverd. Tela, 8:2273 (1966); 7:2109 (1966).

63. A. Karo and J. Hardy, Phys. Rev., 129:2024 (1963).

64. C. Walker and R. Pohl, Phys. Rev. 131:1433 (1963).

65. G. S. Zavt, Trudy Inst. Fiz. Astron. Akad. Nauk Eston. SSR, No. 29:95 (1964).

66. K. K. Rebane, A. A. Rentel', and O. I. Sil'd, Inzh.-Fiz. Zh., 2:60 (1959).

67. D. L. Dexter, Nuovo Cimento, Vol. 7, ser. X, Suppl. 2:245 (1958).

68. K. K. Rebane, A. P. Purga, O. I. Sil'd, and V. V. Khizhnyakov, Trudy Inst. Fiz. Astron. Akad. Nauk Eston. SSR, No. 14:31 (1961).

69. K. K. Rebane and O. I. Sil'd, Optika i Spektr., 13:465 (1962).

70. L. Schiff, Quantum Mechanics, McGraw-Hill, New York (1956).

71. H. Lorenz, Z. Phys., 46:558 (1928).

72. K. K. Rebane and V. V. Khizhnyakov, Optika i Spektr., 14:362 (1963).

73. R. A. Preém, Trudy Inst. Fiz. Astron. Akad. Nauk Eston. SSR, No. 16:57 (1961).

74. S. L. Kukushkin, Zh. Éksp. Teor. Fiz., 44:703 (1963); Fiz. Tverd. Tela, 5:2170 (1963).

75. E. D. Trifonov and V. I. Tamarchenko, Vestnik Leningrad State Univ., 16:21 (1965).

76. V. A. Loorits and K. K. Rebane, Trudy Inst. Fiz. Astron. Akad. Nauk Eston. SSR, No. 32:3 (1967).

77. L. Van Hove, Phys. Rev., 89:1189 (1953).

78. I. Phillips, Phys. Rev., 104:1263 (1956).

79. N. N. Kristofel', Trudy Inst. Fiz. Astron. Akad. Nauk Eston. SSR, No. 26:13 (1964).

80. N. N. Kristofel', Fiz. Tverd. Tela, 7:2519 (1965).

81. K. K. Rebane and L. A. Rebane, Izv. Akad. Nauk Eston. SSR, ser. fiz.-mat. i tekhn. nauk, 14:309 (1965).

82. J. J. Rolfe, J. Chem. Phys., 40:1664 (1964).

83. K. K. Rebane, L. A. Rebane, and O. I. Sil'd, Preprints of the International Conference on Luminescence, 5:115, Budapest (1966).
84. M. Lax, J. Chem. Phys., 20:1752 (1952).
85. K. K. Rebane, Optika i Spektr., 9:557 (1960).
86. K. K. Rebane, A. P. Purga, O. I. Sil'd, and V. V. Khizhnyakov, Trudy Inst. Fiz. Astron. Akad. Nauk Eston. SSR, No. 14:31 (1961).
87. Ch. B. Lushchik, N. E. Lushchik, and I. A. Muuga, Trudy Inst. Fiz. Astron. Akad. Nauk Eston. SSR, No. 23:22 (1963).
88. J. J. Markham, Revs. Mod. Phys., 31:956 (1959).
89. Yu. E. Perlin, Usp. Fiz. Nauk, 80:553 (1963).
90. W. Heitler, The Quantum Theory of Radiation, Oxford Univ. Press, New York (1954).
91. D. A. Patterson and C. C. Klick, Phys. Rev., 105:401 (1957).
92. H. Frauenfelder, The Mössbauer Effect, W. A. Benjamin, New York (1961).
93. V. I. Gol'danskii, The Mössbauer Effect and its Application in Chemistry, Izd. Akad. Nauk SSSR (1963).
94. Encyclopedic Physics Dictionary, Izd. Sovetskaya Entsiklopediya (1963).
95. R. Silsbee, Phys. Rev., 128:1726 (1962).
96. K. K. Rebane, Optika i Spektr., 16:594 (1964).
97. K. K. Rebane and O. I. Sil'd, Trudy Inst. Fiz. Astron. Akad. Nauk Eston. SSR, No. 26:25 (1964).
98. K. K. Rebane and V. V. Khizhnyakov, Crystal Spectroscopy, Proceedings of the Symposium on the Spectroscopy of Crystals Containing Rare-Earth Elements and Iron Group Elements, February 3-6, 1965, Moscow, Izd. Nauka (1966), p. 33.
99. H. J. Lipkin, Ann. Phys. (NY), 9:332 (1960); 18:182 (1962).
100. V. V. Khizhnyakov, Trudy Inst. Fiz. Astron. Akad. Nauk Eston. SSR, No. 32:16 (1966).
101. R. Dicke, Phys. Rev., 89:472 (1953); P. Wittke and R. H. Dicke, Phys. Rev., 103:620 (1956).
102. V. V. Khizhnyakov, Izv. Akad. Nauk Eston. SSR, ser. fiz.-mat. tekhn. nauk, 17 (1968).
103. V. A. Ditkin and A. P. Prudnikov, Integral Transforms and Operational Calculus, (Reference mathematical library, ed. L. A. Lyusternik and A. P. Yampol'skii), Fizmatgiz (1961), p. 17.
104. L. D. Landau and E. M. Lifshits, Quantum Mechanics, Fizmatgiz (1963).
105. V. V. Khizhnyakov, Trudy Inst. Fiz. Astron. Akad. Nauk Eston. SSR, No. 20:67 (1963).
106. K. K. Rebane, O. I. Sil'd, and I. Yu. Tekhver, Trudy Inst. Fiz. Astron. Akad. Nauk Eston. SSR, No. 27:23 (1964).
107. K. K. Rebane, A. P. Purga, O. I. Sil'd, and V. V. Khizhnyakov, Trudy Inst. Fiz. Astron. Akad. Nauk Eston. SSR, No. 14:48 (1961).
108. M. Forro, Z. Phys., 56:534 (1929); 58:613 (1929).
109. W. Martiensen, Proc. Internat. Conf. Semicond. Phys., Czech. Academy of Sciences, Prague (1960), p. 760.
110. R. A. Avarmaa, K. K. Rebane, and V. V. Khizhnyakov, Trudy Inst. Fiz. Astron. Akad. Nauk Eston. SSR, No. 29:76 (1964).

111. D. E. McCumber, Phys. Rev., 134:A299 (1964).

112. V. G. Fedoseev and V. V. Khizhnyakov, Trudy Inst. Fiz. Astron. Akad. Nauk Eston. SSR, No. 29:90 (1964).

113. G. S. Zavt and N. N. Kristofel', Optika i Spektr., 13:229 (1962).

114. Yu. Kagan, Zh. Éksp. Teor. Fiz., 47:366 (1964).

115. E. Hanamura and T. Inui, J. Phys. Soc. Japan, 18:690 (1963).

116. R. J. Elliott, Lattice Dynamics, Proc. International Conf., Copenhagen, ed. R. F. Wallis (1963); J. Phys. Chem. Solids, Suppl. 1:459 (1965).

117. W. Hayes, G. D. Jones, R. J. Elliott, and C. T. Sennett, Lattice Dynamics, Proc. International Conf., Copenhagen, ed. R. F. Wallis (1963); J. Phys. Chem. Solids, Suppl. 1:475 (1965).

118. M. A. Ivanov, L. B. Kvashnina, and M. A. Krivoglaz, Fiz. Tverd. Tela, 7:2047 (1965).

119. M. A. Ivanov, M. A. Krivoglaz, D. N. Mirlin, and I. I. Reshina, Fiz. Tverd. Tela, 8:192 (1966).

120. I. P. Ipatova and A. A. Klochikhin, Zh. Éksp. Teor. Fiz., 50:1601 (1966).

121. R. A. Avarmaa and L. A. Rebane, Izv. Akad. Nauk Eston. SSR, ser. fiz.-mat. i tekhn. nauk, 18:1 (1969).

122. L. Rebane and T. Saar, Izv. Akad. Nauk Eston. SSR, ser. fiz.-mat. i tekhn. nauk, 15:297 (1966).

123. L. A. Rebane, in: Molecular Luminescence Centers O_2^- and S_2^- in Alkali Halide Crystals, Trudy Inst. Fiz. Astron. Akad. Nauk Eston. SSR, No. 37:14 (1968).

124. K. K. Rebane and O. I. Sil'd, Izv. Akad. Nauk Eston. SSR, ser. fiz.-mat. i tekhn. nauk, 15:299 (1966).

125. C. C. Klick, Phys. Rev., 85:154 (1952).

126. Ch. B. Lushchik, N. E. Lushchik, and K. K. Shvarts, Optika i Spektr., 9:215 (1960).

127. M. A. Krivoglaz, Dissertation, Kiev (1954).

128. Yu. Kagan, Introduction to the collection "The Mössbauer Effect," IL (1962).

129. R. Barrie and K. Nishikawa, Can. J. Phys., 41:1135, 1823 (1963).

130. M. A. Krivoglaz, Ukr. Fiz. Zh., 11:1331 (1964).

131. B. Z. Malkin, Fiz. Tverd. Tela, 5:3088 (1963).

132. A. L. Schawlow, Advances in Quantum Electronics, ed. J. R. Singer, Columbia Univ. Press, New York (1961).

133. W. Low, Advances in Quantum Electronics, ed. J. R. Singer, Columbia Univ. Press, New York (1961).

134. E. B. Aleksandrov, Optika i Spektr., 14:436 (1963); Trudy Komissii po Spektroskopy Akad. Nauk SSSR, No. 1:301, 308 (1964).

135. A. P. Purga, Izv. Akad. Nauk Eston. SSR, ser. fiz.-mat. i tekhn. nauk, 15:334 (1966).

136. F. D. Klement, Izv. Akad. Nauk SSSR, ser. fiz., 15:651 (1951).

137. J. J. Hopfield, D. G. Thomas, and M. Gershenzon, Phys. Rev. Letters, 10:162 (1963).

138. A. A. Kaplyanskii, Optika i Spektr., 7:677, 683 (1959); 10:165 (1961); Proc. International Conf. Semicond. Phys., Prague (1960), p. 356.

139. G. M. Svishchev, Optika i Spektr., 18:614 (1965).

140. K. K. Rebane and O. I. Sil'd, Optika i Spektr., 23:414 (1967).

141. L. A. Rebane, Izv. Akad. Nauk Eston. SSR, ser. fiz.-mat. i tekhn. nauk, 15:301 (1966).

142. R. D. Kirk, J. H. Schulman, and H. B. Rosenstock, Solid State Communications, 3:235 (1965).

143. V. L. Broude and S. M. Kochubei, Fiz. Tverd. Tela, 6:354 (1964).

144. K. K. Rebane, Crystal Spectroscopy, Proceedings of the Symposium on the Spectroscopy of Crystals Containing Rare-Earth and Iron Group Elements, February 3-6, 1965, Moscow, Izd. Nauka (1966), p. 21.

145. C. L. Tang, H. Statz, G. A. deMars, and D. T. Wilson, Phys. Rev., 136:A1 (1964).

146. V. Khizhnyakov (Hizhnyakov) and I. Tekhver (Tehver), Phys. Stat. Solidi, 21:755

147. W. Voigt and S. B. Bayer, Akad. Wiss., 603 (1912).

148. S. É. Frish, Optical Spectra of Atoms, Fizmatgiz (1963), p. 485.

149. D. W. Posener, Australian J. Phys., 12:184 (1959).

150. W. M. Yen, W. C. Scott, and A. L. Schawlow, Phys. Rev., 136:A271 (1964).

151. N. M. Galaktionova, V. F. Egorova, V. S. Zubkova, and A. A. Mak, Optika i Spektr., 22:68 (1967).

152. D. N. Mirlin and I. I. Reshina, Fiz. Tverd. Tela 5:3352 (1963) Fiz. Tverd. Tela, 6:945 (1964) Fiz. Tverd. Tela, 6:3078 (1964).

153. B. Fritz, U. Gross, and D. Bäuerle, Phys. Stat. Solidi, 11:231 (1965).

154. B. Fritz, Lattice Dynamics, Proc. International Conf., Copenhagen, ed. R. F. Wallis (1963); J. Phys. Chem. Solids, Suppl. 1:485 (1965).

155. T. Timusk and M. V. Klein, Phys. Rev., 141:664 (1966).

1 6. Cheuk K. Chau, Miles V. Klein, and Brent Wedding, Phys. Rev. Letters, 17:521 (1966).

157. R. Metselaar and J. van der Elsken, Phys. Rev., 165:359 (1968).

158. M. A. Cundill and W. F. Sherman, Phys. Rev., 168:1007 (1968).

159. A. A. Maradudin, "Theoretical and Experimental Aspects of the Effects of Point Defects and Disorder on the Vibrations of Crystals - 2," Solid State Physics, ed. Seitz and Turnbull, 19:2 (1966).

160. R. Loudon, Proc. Phys. Soc., 84:379 (1964).

161. T. Gethins, T. Timusk, and E. J. Woll, Phys. Rev., 157:744 (1967).

162. Nguyen X. Xinh, Phys. Rev., 163:896 (1967).

163. John B. Page and B. G. Dick, Phys. Rev., 163:910 (1967).

164. R. Loudon, Proc. Int. Conf. on Light-Scattering Spectra of Solids, New York (1968) (in preparation).

165. O. Sil'd, Izv. Akad. Nauk Eston. SSSR, Fiz. i mat., 17(2):203 (1968).

166. Proceedings of the International Conference on Light Scattering Spectra of Solids, New York (1968) (in preparation).

167. G. Placzek, Rayleigh-Streuung und Raman-Effekt, in: E. Marx, Handbuch der Radiologie, Bd. VI, Leipzig, Teil II (1934), s. 205.

168. I. I. Kondilenko, P. A. Korotkov, and V. L. Strizhevskii, Optika i Spektr., 9:26 (1960).

169. K. Rebane, V. Khizhnyakov, and I. Tekhver, Izv. Akad. Nauk Eston. SSR, ser. fiz.-mat. i tekhn. nauk, 16:207 (1967).

170. I. Tekhver and V. Khizhnyakov, Izv. Akad. Nauk Eston. SSR, ser. fiz.-mat. i tekhn. nauk, 15:9 (1966).
171. B. I. Stepanov and P. A. Apanasevich, Izv. Akad. Nauk SSSR, ser. fiz., 22:1380 (1958).
172. P. A. Apanasevich, Trudy Inst. Fiz. Mat. Akad. Nauk Beloruss. SSR, No. 3:72 (1959); No. 3:187 (1959).
173. I. Tekhver, Izv. Akad. Nauk Eston. SSR, ser. fiz.-mat. i tekhn. nauk, 17:235 (1968).
174. I. Tekhver, Izv. Akad. Nauk Eston. SSR, ser. fiz.-mat. i tekhn. nauk, 15:603 (1966).
175. K. Poiker and E. D. Trifonov, Fiz. Tverd. Tela, 10:1705 (1968).
176. L. N. Ovander, Optika i Spektr., 3:221 (1957); 4:555 (1958); 5:10 (1958).
177. Yu. E. Perlin, Fiz. Tverd. Tela, 2:1915 (1960).
178. I. P. Dzyub and A. F. Lubchenko, Fiz. Tverd. Tela, 3:3602 (1961).
179. A. F. Lubchenko and B. M. Pavlik, Fiz. Tverd. Tela, 5:1714 (1963).
180. Yu. E. Perlin, Yu. B. Rozenfel'd, Uch. Zapisk. Kishinev Gosuniversiteta, 75 :1 (1964).
181. E. D. Trifonov and K. Poiker, Fiz. Tverd. Tela, 7:2345 (1965).
182. A. F. Lubchenko, Phys. Stat. Solidi, 9:879 (1965).
183. K. Rebane and P. Saari, Izv. Akad. Nauk Eston. SSR, ser. fiz.-mat. i tekhn. nauk, 17:241 (1968).
184. N. A. Borisevich, Excited States of Complex Molecules in the Gas Phase, Izd. Nauka i Tekhnika, Minsk (1967).
185. K. Poiker and E. D. Trifonov, Fiz. Tverd. Tela, 8:2039 (1966); E. D. Trifonov and K. Peuker, J. Phys., 26:738 (1965).
186. K. K. Rebane and O. I. Sil'd, Trudy Inst. Fiz. Astron. Akad. Nauk Eston. SSR, No. 12:3 (1960).
187. F. E. Williams and H. Eyring, J. Chem. Phys., 15:289 (1947).
188. F. E. Williams, J. Chem. Phys., 19:457 (1951); F. E. Williams and M. H. Hebb, Phys. Rev., 84:1181 (1951).
189. C. Klick, Phys. Rev., 85:154, 723 (1952); J. Phys. Chem., 57:776 (1953).
190. N. N. Kristofel' and G. S. Zavt, Optika i Spektr., 20:669 (1966).
191. M. A. El'yashevich, M. V. Vol'kenshtein, and B. I. Stepanov, Molecular Vibrations, Gostekhizdat (1949).
192. L. A. Gribov, Theory of the Intensities in the Infrared Spectra of Polyatomic Molecules, Izd. Akad. Nauk SSSR (1963).
193. S. L. Mayants, Theory and Calculation of Molecular Vibrations, Izd. Akad. Nauk SSSR (1960).
194. V. G. Nevzglyadov, Theoretical Mechanics, Fizmatgiz (1959).
195. V. I. Smirnov, A Course of Higher Mathematics, Vol. 3, part I, Fizmatgiz (1959).
196. A. P. Purga, Trudy Inst. Fiz. Astron. Akad. Nauk Eston. SSR, No. 27:57 (1964).
197. N. S. Krylov and V. A. Fok, Zh. Eksp. Teor. Fiz., 17:93 (1947).
198. M. I. Podgoretskii and O. A. Khrustalev, Usp. Fiz. Nauk, 81:217 (1963).
199. B. A. Bryukhanov, N. N. Delyagin, and Yu. Kagan, Zh. Éksp. Teor. Fiz., 45:1372 (1963).
200. L. I. Vidro and B. I. Stepanov, Trudy GOI, 23(139):24 (1953).

201. I. S. Gradshtein and I. M. Ryzhik, Tables of Integrals, Sums, Series, and Products, Fizmatgiz (1962), p. 973.
202. B. Szigeti, Trans. Faraday Soc., 45:155 (1949).
203. J. R. Hardy and A. M. Karo, Phil. Mag., 5:859 (1960).
204. M. Hass, Phys. Rev., 119:633 (1960).
205. A. D. B. Woods, W. Cochran, and B. N. Brockhouse, Phys. Rev., 119:980 (1960).
206. K. S. Krishnan and S. K. Roy, Proc. Roy. Soc., A207:447 (1951).
207. A. D. B. Woods, B. N. Brockhouse, and R. A. Cowley, Phys. Rev., 131:1025 (1963).
208. A. I. Stekhanov and M. B. Éliashberg, Fiz. Tverd. Tela, 2:2354 (1960).
209. W. Kaiser, W. G. Spitzer, R. H. Kaiser, and L. E. Howarth, Phys. Rev., 127:1950 (1962).
210. D. Gribier, B. Farkolex, and B. Yacrot, Phys. Letters, 1:187 (1962).
211. G. F. Imbusch, W. M. Yen., A. L. Schawlow, D. McCumber, and M. D. Surge, Phys. Rev. 133:A1029 (1964).
212. V. V. Gnedenko, A Course of Probability Theory, Gostekhizdat (1954).
213. G. Kramer, Mathematical Methods of Statistics [Russian translation] IL (1948).
214. A. P. Purga, Trudy Inst. Fiz. Astron. Akad. Nauk Eston. SSR, No. 25:109 (1964).
215. V. F. Bratsev, Doklady Akad. Nauk SSSR, 160:570 (1965).
216. S. T. Epstein, J. Chem. Phys., 44:836 (1966).
217. H. Hellmann, Einführung in die Quanten Chemie, Leipzig (1937), p. 285.
218. R. P. Feynman, Phys. Rev., 56:340 (1939).
219. T. Berlin, J. Chem. Phys., 19:208 (1951).
220. W. L. Clinton, J. Chem. Phys., 33:1603 (1960).
221. W. L. Clinton and W. C. Hamilton, Revs. Mod. Phys., 32:422 (1960).
222. J. C. Slater, Electronic Structure of Molecules, Quantum Theory of Molecules and Solids, Vol. I, McGraw-Hill, New York—London (1963).
223. W. L. Clinton and B. Rice, J. Chem. Phys., 29:445 (1958).

Supplementary References

1. A. A. Maradudin, E. Montroll, and J. Weiss, Lattice Dynamics in the Harmonic Approximation, Academic Press, New York (1963).
2. D. B. Fitchen, "Zero-Phonon Transitions," in: Physics of Color Centers, ed. W. B. Fowler, Academic Press, New York (1968).
3. P. Görlich, H. Karras, G. Kötitz, and R. Lehman, "Spectroscopic Properties of Activated Laser Crystals," Phys. Stat. Solidi, 5:437 (1964); 6:277 (1964); 8:385 (1965).
4. A. E. Hughes, "Isotope Shifts for Zero-Phonon and Phonon-Assisted Transitions in Lithium Fluoride," Proc. Phys. Soc., 88:449 (1966).
5. A. F. Lubchenko and S. I. Dudkin, "The Two-Particle Green's Function Method in the Theory of Light Absorption by Impurity Centers in Solids," Phys. Stat. Solidi, 14:227 (1966).
6. G. F. Imbusch, W. M. Yen, A. L. Schawlow, G. E. Devlin, and J. P. Remeika, "Isotope Shift in the R Lines of Chromium in Ruby and MgO," Phys. Rev., 136:A481 (1964).
7. A. A. Maradudin, "Theoretical and Experimental Aspects of the Effects of Point Defects and Disorder on the Vibrations of Crystals," in: Solid State Physics, ed. F. Seitz and D. Turnbull, Vol. 18, pp. 273-420; Vol. 19, pp. 1-134, Academic Press, New York (1966).
8. A. A. Kaplyanskii, "Line Luminescence of Lithium Fluoride Crystals Irradiated with X-rays," Optika i Spektr., 6:424 (1959). Zero-Phonon Line in the Luminescence Spectrum.
9. E. D. Trifonov, "Features of the Band Shape in Optical Spectra of Crystals Containing Impurities and in the Mössbauer Effect," Fiz. Tverd. Tela, 6:462 (1964). Discusses the relation between the structure in the density of vibrational frequencies and in the vibrational structure of the spectrum.
10. P. P. Feofilov, "Orientation of Line Luminescence Centers in LiF Crystals Irradiated with X-rays," Optika i Spektr., 6:426 (1959).
11. A. A. Kaplyanskii and P. P. Feofilov, "Low-Temperature Spectra of Divalent Samarium in Alkali Halide Single Crystals," Optika i Spektr., 16:264 (1964).
12. A. E. Hughes, "Zero-Phonon Transitions and Vibrational Structure," J. Phys., 28:8-9 Supplement, 55-65, 1967 (1968).

13. Yu. E. Perlin, "Optical Transitions of Degenerate Localized Levels," Fiz. Tverd. Tela, 10(7):1941 (1968).

14. Yu. B. Rosenfeld, V. G. Vekhter, and B. S. Tsukerblat, "Optical Bands in Electron-Vibrational Systems with Degeneracy," Phys. Stat. Solidi, 28:2, K179 (1968).

15. H. R. Zeller, R. T. Shuey, and W. Känzig, "The Molecular Character of the O_2^- Center in Alkali Halides," J. Phys. 28, Nos. 8-9, C4-81 (1967).

16. M. Wagner, "Lattice Vibrations in Crystals with Molecular Impurity Centers," Phys. Rev., 131(6):2520 (1963).

17. M. Wagner, "Resonance Scattering of Phonons by Molecular Impurity Centers," Phys. Rev., 133(3A):A750 (1964).

18. R. I. Personov and V. V. Solodunov, "The Temperature Dependence of the Linewidth in Quasiline Spectra of 1,12 Benzoperylene," Optika i Spektr., 23:590 (1967). "Interference Measurements of the Linewidth in the Quasiline Spectrum of 1,12 Benzoperylene at 4 K," Optika i Spektr., 24:142 (1968).

19. Yu. E. Perlin and L. S. Kharchenko, "Theory of Magnetooptical Effects of Localized Electrons," Scientific Papers of the Kishinev State University, 90:3 (1967).

20. A. D. Matthew and A. Hart-Davis, "Theory of Isotopic Effects in the Electron-Phonon Interaction," Phys. Rev., 168:936 (1968).

21. A. I. Stekhanov and M. B. Eliashberg, "Observation of Localized Vibrations in Raman Scattering from a Crystal of KCl with Lithium Impurities," Fiz. Tverd. Tela, 5:2985 (1963); "Raman Scattering of Light by Quasilocalized Vibrations," Fiz. Tverd. Tela, 6:3397 (1964).

22. J. M. Worlock and S. P. S. Porto, "Raman Scattering by F Centers," Phys. Rev. Letters, 15:697 (1965).

23. A. I. Stekhanov and T. I. Maksimova, "Raman Scattering of Light by Quasilocalized Vibrations near Na, Cs, and Rb Impurities in KCl," Fiz. Tverd. Tela, 8:924 (1966).

24. C. J. Buchenauer, D. B. Fitchen, and J. B. Page, in: "Proceedings of the International Conference on Light Scattering Spectra of Solids," New York (1968). (in preparation); Raman Spectra of F Centers."

25. P. S. Pershan and W. B. Lacina, in: "Proceedings of the International Conference on Light Scattering Spectra of Solids," New York (1968). (in preparation); "Raman Scattering from Mixed Crystals and Point Defects."

26. D. A. Kleinman, "Raman Effect in the F Center," Phys. Rev., 134:A423 (1964).

27. Nguen X. Xinh, A. A. Maradudin, and R. A. Coldwell-Horsfall, "Impurity-Induced First-Order Raman Scattering of Light by Alkali Halide Crystals," J. Phys., 26:717 (1965).

28. L. É. Gurevich, I. P. Ipatova, and A. A. Klochikhin, "Raman Scattering of Light in Cubic Ionic Crystals Containing Impurities," Fiz. Tverd. Tela, 8:3260 (1966).

29. C. H. Henry, "Analysis of Raman Scattering by F Centers," Phys. Rev., 152:699 (1966).

30. G. Benedek and G. F. Nardelli, "Raman Scattering by Color Centers," Phys. Rev., 154:872 (1967).

31. M. Ashkin, "Raman Scattering of Light from H^- Centers in CaF_2," J. Phys., 26:709 (1965).

32. T. Timusk and M. Buchanan, "One-Phonon Sideband of Sm^{++} in KBr," Phys. Rev., 164:345 (1967).
33. P. G. Klemens, "Anharmonic Attenuation of Localized Lattice Vibrations," Phys. Rev., 122:443 (1961).
34. A. S. Davydov, "Theory of Urbach's Rule, " Phys. Stat. Solidi, 26:2 (1968).
35. M. A. Nusimovici and J. L. Birman, "Sum Rule for Lattice Vibrations of the Wurtzite Structure," J. Phys. Chem. Solids, 27:701 (1966).
36. M. A. Nusimovici and J. L. Birman, "Lattice Dynamics of Wurtzite: CdS," Phys. Rev., 156:925 (1967).
37. É. I. Rashba, "Theory of Vibronic Spectra of Molecular Crystals," Zh. Éksp. Teor. Fiz., 50:1064 (1966).

Index